Healing Light

Energy Medicine of the Future

by

Dr. Larry Lytle

authorHOUSE™

1663 LIBERTY DRIVE, SUITE 200
BLOOMINGTON, INDIANA 47403
(800) 839-8640
WWW.AUTHORHOUSE.COM

First published by AuthorHouse 12/16/04

ISBN: 1-4208-0200-3 (sc)

Library of Congress Control Number: 2004098781

Printed in the United States of America
Bloomington, Indiana
This book is printed on acid-free paper.

Dedication

This book is dedicated to all seekers of better health!

Special Thanks to those that assisted me; Dr Morton Walker, Dr Li-Chuan Chen and his wife Demerie, my son Kip, my companion Fredretta, and my friends and colleagues who have contributed to my knowledge and understanding.

Foreword

As the editor of a health magazine (*Integrative Health & Self Healing*, www.integrativehealthandhealing.com), I have come across quite a few healing devices. Most of them have failed to make a broad impact because they are designed mainly for health professionals. Since I advocate self-healing and empowerment, especially in light of the looming health crisis in this country, I have special affinity for treatments and devices, such as the Q Series low level lasers, that are user friendly and cost effective.

The medical crisis that we face today includes not only our inability to deal with a variety of chronic degenerative diseases, but also the daunting reality that the cost of medicine will be out of reach for many people in the future. Millions are already uninsured. As the costs continue to rise, many more will join this medical underclass. At the same time, many people are gravitating towards complementary and alternative medicine (CAM), even though it is rarely insured, and they must pay for it "out of pocket." But if the economy worsens, fewer people will be able to afford CAM. Moreover, the federal regulations stipulate that tax deferred health saving accounts cannot be applied to the cost of dietary supplements.

The current situation is not encouraging. If we think politicians will come up with a solution, let's think twice. The prescription drug legislation, supposedly put into place to help senior citizens, actually serves to enrich the pharmaceutical companies while it increases the government deficit. I can't imagine new legislation that will be able to take care of our health without adversely affecting the economy. Therefore, I would rather bet my own money on effective healing tools that I can use at home.

A well thought out health care system should integrate methods for helping people achieve good nourishment, detoxification and regeneration. Dr. Larry Lytle's vision, centered on low level laser, does just this. He sees the photon (the smallest particle of light) as a kind of nutrient. Dr. Lytle's Q Series, 606 laser enhancer, designed for "laser puncture" (a way of doing acupuncture using laser light) helps in the processes of detoxification and regeneration. Dr. Lytle also has two other health products that complement the benefits of low level laser therapy. Belly Gelly®, which helps maintain the colon, supports all three above-mentioned important functions. His Miracle Bite Tabs help heal "dental distress syndrome" and thus improve over-all health. These form a simple and very compact home health care system for prevention and improvement of general health. Of course, for some acute, traumatic and unrelenting diseases, one must rely on physicians and other health professionals for help.

Dr. Lytle emphasizes the elimination of structural problems and nervous imbalances caused by what he calls, "faulty proprioception to the brain." He is a good student of Dr. A. C. Fonder, who was highly praised for his concept of "dental distress syndrome" by Dr. Hans Seleye, the most famous scientist in the field of "stress." Dr. Lytle's Miracle Bite Tabs, a kind of tooth guard easily made in one's own kitchen, alleviate faulty proprioception from a bad bite, and provide a good short-term solution for this serious problem.

I have had the good fortune to discuss "dental distress syndrome" with Dr. Lytle in depth. As a trained toxicologist, I am aware of all the dangers of dental amalgams, galvanism, toxic dental materials, root canal teeth, periodontal disease, improper tooth extraction and bad orthodontic work. However, Dr. Lytle educated me about the consequences of a bad bite and how it can slowly cause serious degenerative disease without being recognized as doing so. I was convinced! I promptly began using his Miracle Bite Tabs and lasers. To my surprise, I found my body respond with five weeks of detoxification in the throat area. For me, that was astounding. I have been working on detoxifying my body for a long time and considered myself relatively "clean." His work, apparently, was able to cleanse a part of me that was not responding to the detoxification procedures that I was already doing.

Although an expert in the professional fields of dentistry, nutrition and neurology, and an inventor of the highest order, Dr. Lytle is still firmly rooted in his past, as a farm boy in rural South Dakota. He is open, generous and caring. When I took Dr. Lytle's weekend workshop on laser, he invited me to spend some personal time with him and his companion at his home in Rapid City, South Dakota. He shared his knowledge without reservation. He could easily retire and live comfortably at his age, but he chooses to teach, help people improve their health, and learn to take care of themselves. He simply cares about people. Moreover, I observed that Dr. Lytle's speech and actions are consistent. He uses his own products and takes good care of himself and his family. This is quite rare. I have seen many healers and doctors who "don't have time" to take care of themselves.

Dr. Lytle's book, *Healing Light*, presents information on the art and science of low level laser and also describes a home health care system. I highly recommend it for all readers

Li-Chuan Chen, Ph.D.

Bethesda, Maryland

August 16, 2004

Disclaimer and Disavowal of Responsibility

This book has been written and published strictly for informational purposes and should not be substituted for consultation with your own health care professional (medical doctor, osteopath, dentist, naturopath, chiropractor, acupuncturist or nutritionist).

The information imparted here is not the practice of medicine, dentistry, chiropractic or any other health care science. The author, consultants and publisher provide educational material only.

If information gleaned here raises questions about your own or a loved one's health and well being, please consult a medical professional. The author is **NOT** a medical doctor.

Unless otherwise indicated in the footnotes, the patients' identities, including their occupation and place of residence, have been changed; their original statements have also been paraphrased. However, the references to clinical studies, research scientists and holistic health care professionals are actual.

The author, publisher, consultants, editorial contributors, cited organizations and product suppliers disavow all responsibility for any application of the information provided in this book. The use of this information for practice, procedures, diagnostic techniques and/or medical devices is the responsibility of the reader and all interested parties.

Thank you

Healing Light - Energy Medicine of the Future

Chapter 1   Introduction

1.1   Meet the Author

Any man can write words, but it is more important to know what sort of man wrote them.

I may be the only person you have met who was delivered by his father in a sod house. I was born in 1935, during the great depression, on the border of the Pine Ridge Indian Reservation in South Dakota.

Fig 1

I do not remember my first year of life, but my sisters referred to it as "a living hell." My mother, father, two brothers and four sisters shared a tiny shanty, a sod house with a dirt floor, log and dirt chinked sides, and a sod roof. It was heated with gathered wood and cow chips (Figure 1). My father, one of the original and successful fat cattle feeders, raised, fattened and

1

delivered beef to the stock yards in Omaha for shipment to the eastern cities. But when the Great Depression hit, he was forced, humbly, to ask the Government for shells to shoot his 100 or so remaining cattle to prevent their death by starvation.

Mother Nature was harsh. It was dry year after year. The streambeds were baked and the grass shriveled. The trees went dormant as hot winds blew. Precious topsoil misted into clouds of dust that drifted like snow, covering fence lines and even buildings. Life was tough, really tough, with seven kids to feed, clothe and educate!

Despite this hardship, our family remained sturdy. I remember preparing my acceptance speech for the Distinguished Service Award which I received from Chadron State College, my *alma mater.* When asking myself, "Why am I here? Who should I thank?" the answer was clear. I expressed appreciation to my remarkable parents! My mom, a "work-a-holic," raised a family of seven with no modern facilities, no money, and only food she could grow and preserve. If Mom couldn't get her work done in 18 hours, she worked 22. If she couldn't take care of her family in a 22-hour day, she worked faster. My dad was a man of wisdom, with a "knowingness" of the energy of the Universe. The Great Depression slowed him down but did not get him down. Even the hail that wiped out his crops did not beat him. The grasshoppers darkened the sun and reduced his 6-foot tall green fields to 6-inch stubble in a few short days. Yet he continued. Mom and Dad were tough people with great perseverance. They had principles, integrity and loyalty. Though not highly educated, they were smart. Mom had one year of Business College and dad might have completed the third grade.

I have more degrees than my father. Some day I hope to have the wisdom he had.

I attended part of my grade school in the small town of Wasta, SD, a town of under 100 people. I finished grade school at Wall, the home of the famous Wall Drug Store. Our farm was one mile from the nearest road and there were no buses. Dad always found ways to get us to school, sometimes in a wagon pulled by a team of horses, sometimes in an old car when it could be jump-started by pulling it with the horses. Often we walked but we ALWAYS got to school. It was my parents' top priority for us to have the education they didn't have.

I graduated from Chadron Prep School in 1952 and received some local notoriety in basketball as an All-State player on the Nebraska, Class C, and 1952 undefeated State Champions team. My collegiate studies were at Chadron State College where I also stood out as a basketball player. I made the All-Conference teams and eventually was inducted into the Basketball Hall of Fame. I owe a lot of my success to sports, especially basketball. Following graduation in 1956, I taught high school science and coached basketball at Hot Springs, SD, for two years until I was inducted into the Army. While in the Army I played basketball for Brook Army Medical Center and completed the college courses required for dental school. I graduated from the University of Nebraska, College of Dentistry, in 1964 and proceeded to join my brother in a general dentistry practice in Rapid City, SD.

My practice was successful because I never stopped learning. I studied and improved my knowledge and skill in many facets of general dentistry, including advanced tissue and bone grafting techniques for preserving periodontally-involved teeth, prevention, and mercury replacement with composite materials, and various splint techniques for TMJ/TMD. As an "ole farm boy," I knew that animals died if they lost their teeth. With their permission, I successfully tried many techniques to save patient's teeth. If someone wanted an extraction, I would ask, "Why do you want your tooth pulled?" Then I would educate them about the value of keeping their own teeth. I did many root canal fillings for the price of an extraction.

Preventive Dentistry was my passion. I was an early member of the American Society for Preventive Dentistry and started a state chapter in South Dakota. I lectured and taught lay people and professionals about preventing decay and gum disease. When The American Society of Preventive Dentistry felt that it had achieved its goal of educating dentists and the public, it disbanded. I continued my education by joining the International Academy of Preventive Medicine and served as president for five years.

After incorporating the Academy of Nutrition into our organization and changing our name to the International Academy of Nutrition and Preventive Medicine (IANPM), we became the premier professional organization promoting alternative and complementary care at that time. IANPM had the distinction of publishing the only refereed journal in preventive medicine. During my tenure at IANPM, I had the privilege of learning from great pioneers such as Linus Pauling, Rodger Williams, Emanual Cheraskin and others. It was an exciting time. I became "convinced" that prevention, nutrition and other forms of what we now call "complementary" or "integrative" medicine were very important, and I had a new understanding of the limitations of traditional dentistry. I came up with a slogan that I often used in my lectures, "<u>Once you know something, you cannot unknow it</u>."

My interests led me to study cosmetic dentistry along with my daughter, Dr. Kelly Lytle. We were the first (and I think only) father and daughter team to be accredited by the Academy of Cosmetic Dentistry. Cosmetic Dentistry involves the gum tissue as well as the teeth, and in the early 1990s I learned to use cutting lasers to contour gum tissue. My interest in lasers earned me a category II accreditation with the Academy of Laser Dentistry. The development of laser bonded splint techniques resulted in an invitation to speak at the World Congress on Lasers in Dentistry in Singapore in 1994. I became interested in low level lasers, also referred to as cold lasers, soft lasers or low power (meaning a low power output) lasers. However, when I researched this type of laser I found many inconsistencies. I could not understand how one study showed quite good results while another study of the same type of laser showed poor results. When I used my own cutting (curing) lasers, they consistently worked well. I set out to learn why the world's low level lasers were so inconsistent. My studies, combined with my basic knowledge of energy, propelled me into a new project--- to design a low level laser system that worked consistently.

The body functions as a unit, a system, and part of that system has to do with what I call "proprioception to the brain." My basic knowledge of this topic was gleaned from one of my great teachers, Dr. Al Fonder. Dr. Fonder was one of the first maxial facial surgeons in the military and he restored hundreds of faces blown apart from the ravages of World War II. He learned that the posterior occlusal support, which he restored while working on their faces, improved their entire bodies.[1] In other words, the height of the back teeth influences the relationship of the lower jaw to the skull and thus, affects the entire sympathetic and parasympathetic nervous system. Based upon Fonder's work, neurological research and the early Chinese understanding of meridian energy, I developed the concept of Proprioceptive Feedback to the Brain. To temporarily correct faulty proprioception, I developed proprioceptive structural guides, called "Miracle Bite Tabs," that people can make in their own homes. When proprioceptive guides are combined with low level laser therapy, the results are truly amazing.

One might ask; how does this personal history relate to a book about lasers? I believe that knowing something about an author's real life experience puts his/her work in context.

1.2   Energy Medicine and Healing Light

As a small boy growing up on the Cheyenne River in western South Dakota, I would gaze at the moon and stars. There were too many stars to count and they seemed to extend forever in the clear night air. How far away were they, I wondered? What is the emptiness and vastness of the sky? I imagined cows and faces on the moon as I recited the nursery rhyme, "the cow jumped over the moon." But stars were not just for gazing; as a farm boy I knew how to use the celestial compass to find north --- find the Big Dipper and then follow three handle lengths up the lip and there is the North Star. I knew the Milky Way and Mars long before they were candy bars.

I also wondered where water comes from, why there wasn't more of it. We had a shallow well near the house but that water was "no good" for human or animal consumption. The livestock wouldn't drink it and when we did we got diarrhea. We collected rainwater for drinking but the rains were unpredictable. Most of our water was hauled in a big tank six miles from the town of Wasta (an Indian word for "pure water") and stored in a cistern. About that bad well, if the water was "brackish," too alkaline for consumption, it certainly healed cuts and wounds. Once I cut a deep gash in my leg while whittling with my dad's special, very sharp knife. I was afraid to tell him because I wasn't allowed to use his good knife. I rushed to the brackish well and pumped cool well water on my leg until it stopped bleeding. I often wondered why that cut seemed to heal overnight. Now I know it was the energy.

I was fascinated by Dad's method of locating underground water. He held a V-shaped willow branch, one end in each hand, and scoured the ground with back and forth motions. This is called "dousing" today, but dad used the word "witching." He only "witched" water for our family and his best and most trusted friends because old superstitions still prevailed in his mind. In old England this practice of locating water was considered "witch craft" and could lead to being burned at the stake.

I remember Dad finding water on my sister's farm. Like us, she and her family had been hauling and storing in a cistern for years. Dad surveyed the terrain and using his "willow witch stick" traced water right under the corner the house. Dad instructed my brother and me to dig and assured us that we would find water at twelve feet deep. Sure enough, at twelve feet we found pure water and piped it into the house. His willow branch had traced the water right under the corner of my sister's house. We buried a barrel in the hand dug well, filled it with big rocks to serve as a collection tank, placed a pipe and ran it right into my sister's porch where we installed a pump. She had water in her house for the first time, all because of this mysterious energy Dad called "witching." How did he do it? Years later, I began to work with this mystery of Universal Energy for healing rather than "witching water."

I am still learning along with other scientists and seekers about the vastness of the Universe. I am learning about the Black Hole, the

Zero Point Field and non-local energy. Questions still loom about the conscious mind, the origin of thought and why two people can have the same thought at the same instant thousands of miles apart. We wonder how the birds know when to fly south in the winter and back north in the summer, what takes the swallows back to Capistrano every year on the same day? I understand, in 2002, they were a day late for the first time. Why is that?

Inquiring minds will gradually shed light on these mysteries. In this book I will focus on the mystery of lasers, a form of light and energy healing that will change people's lives and will likely change the practice of medicine.

1.3   Laser, Light and Healing Energy

Many of us have heard of lasers, but more in the context of laser guided bombs or LASIK (for an eye-cutting procedure). This book focuses on the low level laser or "cold" laser, a form of healing light that replenishes lost cellular energy missing in the body due to injury, illness and aging. To present a case, let's consider the following story of Dr. Fritz-Albert Popp's exciting discoveries in energy medicine.

While working at the University of Marburg in Germany during the seventies, Fritz-Albert Popp tried to determine what effect he'd get if he excited the deadly x-ray beam called a Roentgen, with ultraviolet light. He discovered that every carcinogen reacted only to light at the specific wavelength of 380 nanometers. This wavelength is specific for the cellular mechanism of DNA repair, which is blocked by none other than carcinogens. He learned he could repair a 99 per cent damaged cell by illuminating it with a very weak intensity of that same wavelength of light, a phenomenon he called "photo-repair." Today it is called "constructive interference."

Popp and one of his students developed a photomultiplier that could measure light emitted by cells, photon by photon. He went on to show that our own cells produce biophotons that communicate with one another. Indeed, this means we <u>are</u> "light" beings as spiritual seekers have always claimed. What's more, he found that photons in living systems were coherent, as is laser light. In quantum physics, coherence means that these subatomic particles are able to cooperate with each other and are highly interlinked by bands of common electromagnetic fields. Popp proved that coherence establishes communication not only between living cells but within subatomic particles as well. He discovered that more complex organisms emitted fewer biophotons than simpler plants and animals, but both emitted in the range of 200 to 800 nanometers. He also discovered delayed luminescence, a corrective device; if living tissues receive too much light, the excess is rejected. Similarly, when the body is in a stressed state, the emission of biophotons goes up. This is a defense mechanism that is designed to return the body to equilibrium. The healthiest body would be in balance and closest to the Zero Point Field, the most desirable state.[2]

Popp's discoveries explain why some healthy bodies do not respond to low level laser therapy as readily as do sick bodies. It also explains why excess low level laser radiation is ignored by the body. These issues will be analyzed in depth in the following pages, which seek to explain how the amount of energy delivered by low level lasers is sufficient to bring about healing. We will present the positive findings of laser light's photon stimulating effect from double blind studies taking place around the globe. There are many such studies that demonstrate the healing potential of low level lasers. Healing light systems are designed to carry electrons back into the body to replace those lost due to injury, sickness, disease, or the natural process of aging.

Another finding relevant to laser energy came from Herbert Frohlich of the University of Liverpool. He was the first to introduce the idea that a collective vibration was responsible for getting proteins to cooperate with each other to carry out the instructions of DNA. He showed that vibrations in proteins set up a wave communication, thus verifying that biophoton intercellular energy exists. When the molecules begin to vibrate in unison, they reach a high level of coherence and take on certain qualities of quantum mechanics that regulate the body, even via nonlocality. Nonlocality refers to the transfer of energy from a distance ---energy is everywhere instantly.

Nobel Laureate, Albert Szent-Gyorgyi suggested that protein cells act as semiconductors, preserving and passing along energy of electrons that carry information.[3]

Let's briefly return to the story of Popp. A significant breakthrough in Popp's work came when he began thinking of how light produced energy in plants through photosynthesis. He theorized that when we eat broccoli, it is digested into carbon dioxide (CO_2) and water, but what happens to the light from the sun that activated the photosynthesis in the broccoli? It must be stored somehow. When light is taken into the body as photons, it is dissipated and distributed over the entire spectrum of electromagnetic frequencies from the lowest to the highest. This energy becomes the driving force of all the molecules in our body and makes it easier to understand that low level lasers, as a source of photons, affect all molecules in the body. This is also why we are told to eat foods of many different colors, which vary in frequency.

Popp found that molecules in the cells, besides DNA and proteins, would respond to certain frequencies, and that a range of vibrations from photons would cause a variety of frequencies in other molecules of the body. DNA was capable of sending out a large range of frequencies linked to certain functions. Any wave in the Universe can be interrupted, and when this occurs, different frequencies, including those used in low level laser, are created. While much remains to be discovered about which frequencies benefit the body, it is clear that frequency medicine will shape our future.

The hologram further makes the case that we are "beings of light." Dennis Gabor (the father of holography) coined the terms hologram and holography in 1947. The word hologram is derived from the Greek words "holos" meaning whole or complete and "gram" meaning message.[4] The striking feature of the hologram concept is that any dissected piece of information resembles the whole. In fact, this implies that each cell in our body is the same as our whole self. The holographic principle exists in everything.

The original light used for holograms was not very accurate and the images were unclear. Laser light, dramatically different than all other light sources whether man-made or natural, moved holograms into the forefront of science. Laser light was able to do this because it is coherent. This means that the light being emitted by the laser is of the same wavelength, travels at the same speed and is "in phase." The hologram is a recording of the interface of laser light waves bouncing off the object, shown in relation to another coherent laser beam which serves as a reference.

Recent scientific research has provided us with an explanation of the healing power of light. Our knowledge has greatly expanded. We know why light heals and how much is needed. We know, for instance, that jaundice at birth can be reversed by exposure to blue light, and simulated sunlight helps Seasonal Affected Disorders. I have developed a method of dispensing light into the body at the correct power and frequency, just as the sun enters plants in photosynthesis. The following pages discuss how new low level laser technology provides an answer.

1.4   Three Laser Devices for Biostimulation

The *Q1000* is a hand-held laser device about the size of a cordless telephone that operates on rechargeable lithium-ion batteries. Such batteries were selected because the consistent power level they maintain is needed to produce a precise intensity of laser light. Thanks to the miniaturization of modern electronics, the Q1000 contains an onboard circuit and a powerful small processor that controls the eight LEDs and twelve laser diodes that control the emission of seven different wavelengths. These computer controlled diodes form six different soliton waves (explained in chapter 3) that produce a controlled output consistency not found in other lasers. With many laser diodes working together, the Q1000 is capable of emitting a wide range of light wavelengths that are beneficial to the body. The processor controls the power output and time duration and is pre-programmed at the factory with thirty-one frequencies in three different modes. Various combinations promote healing for specific conditions. Moreover, the Q1000 is designed so that new combinations of frequencies can be added when indicated by new research and experience.

The ***660-Enhancer*** is a 50-milliwatt (0.005 of a watt) laser wand that operates at 30 milliwatts and emits a red light wavelength. About the size of an old-fashioned fountain pen, the 660 Enhancer fits comfortably in the hand. It plugs into the Q1000 for its electrical power and control switch. The 660-Enhancer laser produces more joules of energy (6.6 joules in three minutes) than the Q1000 and can be used to stimulate acupuncture points in place of needles, allowing patients to treat themselves noninvasively.

The ***808-Enhancer*** is a 500-milliwatt (one half of a watt) laser wand that operates at 300 milliwatts and emits an infrared (longer than red) wavelength of light. It is the same size as the 660-Enhancer and also plugs into the Q1000 for electrical power and controls. The 808-Enhancer laser is the most powerful of the three lasers and therefore, its laser light penetrates (permeates) deeper into the body. It has been reported by users (and has been verified in clinical research) that the 808-Enhancer laser stimulates bone growth, regenerates cartilage, and improves bone marrow consistency. This is consistent with laboratory research that shows laser light can stimulate bone growth. A laser similar to the 808-Enhancer has been approved by the FDA for treatment of carpal tunnel syndrome.

These three lasers produce important frequencies for healing. The chapters that follow will illustrate their applications in greater detail. Chapter two provides a brief review of the nature of energy. Laser, a form of energy, will be discussed in chapter three. This will lay the foundation for our discussion of low level laser therapy in chapter four. The potential health benefits of low level laser therapy will be described in more detail in chapter five. Chapter six presents the application of low level laser therapy in acupuncture, a modality frequently used in Traditional Chinese Medicine.

Chapters seven and eight discuss beneficial use of low level laser therapy in Dental Distress Syndrome and other dental procedures. The reversal of Dental Distress Syndrome is crucial in dealing with chronic degenerative diseases. Dental Distress Syndrome creates faulty proprioception, which contributes to dystonia, a dysfunction of the autonomic nervous system, which affects almost all bodily functions.

Chapter nine introduces a home health care laser system. Chapter ten presents findings from worldwide low level laser therapy research.

My rich background, resplendent with the mysteries of nature, left me in a state of awe and wonder. Perhaps, that is why I never doubted the energy of the Universe. I have pulled together and condensed information from diverse sources into my book, ***Healing Light***. I hope you enjoy the book and learn how to take charge of your own health through the energy of low level lasers, the true healing light.

Chapter 2   Energetics in a Nutshell

2.1   Historical Understanding of Electromagnetic Energy

Throughout the nineteenth century, scientists experimented and theorized about electricity and magnetism. They had already discovered the intersection between these two types of energy and knew that flowing electricity generates a magnetic field. They were also aware that movement within this field, across the lines of force, generates a flow of electricity. However, despite their familiarity with the theories and experimental evidence, it was difficult to imagine a type of energy that required no medium to flow through. Then some experiments were conducted that left no doubt.

In the 1840s, British scientist Michael Faraday, famous for his intuitive theoretical approach to experiments with electricity and magnetic fields, suggested there might be a link between electromagnetism and light. This intrigued researchers who had seen electric sparks. But what was producing the light?

Then in 1873, another British scientist, James Clerk Maxwell, whose theories were more mathematical, incorporated Faraday's ideas into his own grand hypothesis entitled "Treatise on Electricity and Magnetism." Maxwell's formulas explained many experimental observations about electricity and magnetism. He predicted the discovery of wave radiation that traveled at the same speed as light within an electromagnetic field. He theorized that light is, in fact, electromagnetic. But as yet there was no experimental proof of this.

A decade later, the young German scientist, Heinrich Rudolf Hertz, was able to devise experiments that proved Maxell's hypothesis about the nature of light. Working in a dark room, Hertz detected a spark jump across a gap in a loop of copper wire located within a strong electromagnetic field. The movement through the electromagnetic field had induced an electric current to flow through the copper loop. The current was strong enough for the energy to flow across the small space and complete the circuit. By doing this, Hertz had succeeded in generating invisible radio waves. The visible spark, caused by combustion of the oxygen molecules in the air which were heated by the flowing waves, convinced Hertz that energy was flowing through the gap. Technology progressed swiftly after the Hertz "radio wave" experiment.

Another experiment Hertz conducted with copper wire loop required the attachment of shiny metal plates to each side of the gap. When the loop was connected to a battery nothing happened. When light was shined on one of the plates, it caused energy to flow across the gap, creating an electric current through the entire circuit. Subsequent experiments showed that this also occurred in a vacuum, proving that a medium such as air was not necessary for energy to flow between the metal plates. Laser light, unlike sound, is able to travel through a vacuum as well.

Additional experimentation indicated that different metals react to specific light frequencies. The phenomenon demonstrated by these investigations, which used a beam of light as part of an electronic circuit, was called the "photoelectric effect." Although it could be observed experimentally, the photoelectric effect was not understood until twenty years later. Albert Einstein suggested an explanation in the first of five papers he published in 1905, changing the course of physical science. But before that, one more important event occurred at the end of the nineteenth century.

An Italian inventor by the name of Gugleilmo Marconi demonstrated a "wireless telegraph" that used radio waves to transmit messages through "empty space." In 1896 Marconi patented his device in Great Britain, and in 1901 his wireless signals spanned the Atlantic Ocean!

2.2 Albert Einstein Changes the Course of Physics

Albert Einstein was a mathematical and philosophical genius. Although intrigued by the ideas of other original thinkers who questioned established theories, Einstein preferred to work on his own. In 1905 at the age of twenty-six, while employed by the Swiss Patent Office, Einstein published the first of his five papers on theoretical physics. While theorizing about the nature of light, he managed to explain the photoelectric effect. The young scientist proposed that light energy was transmitted in small amounts that he called "quanta." The quanta would later be renamed "photons."

Einstein identified the characteristic behavior of photons as waves. A photon's wavelength determines the amount of energy it contains. His simple formula for calculating the amount of energy in a photon was wavelength frequency (the number of wavelengths repeated per second or "hertz") multiplied by a tiny number called "Planck's constant" (6.626×10^{-34} developed by German physicist, Max Planck in 1900).

The paper containing Einstein's formula shows: (1) all forms of light energy are photons; (2) all photons contain energy; (3) photons have different wavelengths and hence contain different amounts of energy. Shorter wavelengths have more energy because their frequency is greater.

2.3 Electrons Absorb Photons

Scientists were correct to classify light as electromagnetic radiation. As with magnetism, light and electricity are interrelated. When a photon encounters an electron, the electron absorbs the photon's energy; when sufficient energy is absorbed, the electron moves to another energy threshold. Different wavelengths determine energy content.

In the previously described experiment demonstrating the photoelectric effect conducted by Hertz, a photon hit the surface of metal plate. The energy contained in the photon is absorbed by an electron in one of the atoms of the metal plate. This absorption of energy causes the electron to reach a higher energy threshold. The energized electron is forced to move out of its orbit around the atomic nucleus of the metal atom and flow freely until it is captured by another atom. We call this process "electric current" or "electricity!" Hertz's demonstration that photons can trigger this movement of electrons was an important contribution.

The Hertz experiments also showed that metals respond differently to various frequencies of light. For example, a bright yellow light has no effect on a copper plate. Change the wavelength to ultraviolet (a wavelength shorter than violet and not visible to the human eye) and electrons flow through the copper plate, even if the light is very weak.

Electrons also release excess energy by emitting photons. These photons are then absorbed by other nearby electrons which are traveling at the speed of light and radiate away until they encounter something else that absorbs them.

2.4   Photons Are Small But Powerful

All light is composed of photons. They are so small that about a thousand billion photons of sunlight hit the head of a pin each second. Individual photons cannot be seen by the naked eye but when they travel in groups they become visible. A star that we see as a tiny point of light in the night sky is so distant that only a few hundred photons emitted by it collide with our eyes each second. But that tiny number still allows us to see the star.

Despite the minuscule size of its component photons, light is a powerful force – pure energy with many unique characteristics. Photons have no mass at all and travel faster than any other particle or energy in the universe. Stable in and of themselves, photons combine their energy readily with electrons. Light is in constant motion and does not change its direction unless it is absorbed (by an electron), refracted (the path is bent as the photons pass through a solid object), or reflected (the photons bounce off a solid that they cannot penetrate).

Einstein predicted the path of light could be bent by gravity and could trigger chemical reactions; both concepts were later shown to be true. Hertz' experiments illustrated that photons could trigger an electric current by energizing electrons in, what is called, the "photoelectric effect." Photons can have the same effect, chemically, by supplying the energy for atoms to reconfigure themselves into different combinations. Energy stimulates a chemical reaction in three ways: by means of electricity, heat or light.

2.5   Einstein Redefines the Structure of Reality

Einstein's third scientific treatise, published in 1905, and entitled, "On the Electrodynamics of Moving Bodies," is also known as the "Theory of Relativity." Einstein argues that the speed of light is the only natural phenomenon that never changes. Regardless of its source, light always travels at the same speed. All other things that can be observed or measured in nature, whether large or small (such as sound waves or the rotation of planets around the sun) are relative to the motion and position of the observer. In other words, the actual way something behaves is influenced by, and influences, our perspective, our motion and the way we experience or perceive that it behaves.

Nothing ever stands still. Even when we're sitting very quietly in a chair, the chair is on the surface of the earth, which is spinning on its axis at over 1,000 miles per hour (the earth is about 25,000 miles in circumference and makes one complete revolution in 24 hours). At the same time, the earth is traveling about 67,000 miles per hour in an elliptical orbit around the sun. The sun and the entire solar system revolve around the center of the Milky Way galaxy at a speed estimated to be about 150 miles per second (one-half million miles per hour). Therefore, we're moving continuously. Everything that happens around us is affected by this motion. As we move faster, space is shortened and time is slowed down, a situation that happens to everything relative to our frame of reference, except the speed of light. Only the speed of light stays constant regardless of our motion relative to its source. Light transcends the effects of relative motion.

2.6   Mass = Quantity of Matter

Solid objects have what physicists call "mass," which equals the amount of matter being held together. Laboratory experiments often measure the weight of mass. Although this measurement is a convenient way to calculate the result of an experiment, weight is not a good definition of mass because weight depends on gravity. Things weigh less on the moon because the moon is a smaller solar body than the earth and exerts less gravitational force. The mass of an object is the same on either solar body.

If you and your pet hippopotamus take a trip in your spaceship to an area of space where there is no gravity, you and that hippo will both be weightless. Start playing tag inside the spaceship, however, and you'll find that the hippo can still crush you against the wall when it floats into you at high speed. Its muscular body is denser than yours, and there's more of it. You'll feel the hippo's greater mass overwhelming your lesser mass.

2.7   Einstein Defines the Underlying Structure of Matter

Albert Einstein concluded, in his fifth scientific paper published during that same phenomenal year of 1905, that the scientific terms "energy" and "mass" are only different names for the same thing. He put forth his famous equation $E=mc^2$ (energy equals mass times the speed of light squared).

The following paragraph explains what the letters mean when applied to Einstein's formula: "E" is energy produced in *joules* (a joule is a unit of energy in physics named after the British scientist, James Prescott Joule, for his discovery of the conversion of work into heat). A joule is equivalent to the work done by an electric current of one *ampere* against a resistance of one *ohm* for one second.

The "*m*" is the total mass in kilograms lost (transformed) during the conversion of mass to energy. The "*c*" is the speed of light in meters per second, which is squared (multiplied by itself).

Because "c^2" is a gigantic number, this formula of Einstein's shows that a very small amount of matter contains an enormous amount of stored energy. In other words, the mass of solid objects is actually made of energy.

2.8   Energy is Stored in Solid Objects

Energy of all sorts is stored in solid objects. When the World Trade Center in New York City was being constructed, giant cranes fought the force of earth's gravity and hauled huge steel beams up to 1,353 feet in the air. These steel beams were bolted together like a child's erector set, forming the infrastructure of this monumental building. The potential energy this physical action required was stored in the structure. All the building components and objects inside the building (concrete plate glass, office furniture and equipment, even a coffee mug sitting on a desk) contained this same form of potential energy.

All of this energy was released when the twin towers collapsed following the terrorist attack of September 11, 2001. Huge steel beams crumpled like tin foil. Small objects such as file cabinets and computers disintegrated from compression when the force of earth's gravity reasserted itself on 1.6 million tons of mass. Curiously, many of the plastic identification cards worn by employees of companies using security cardkeys to unlock doors, survived intact from the wreckage. Nobody knows why.

The largest physics experiment in history occurred on July 16, 1945 at 5:29 AM over several hundred miles of flat scrubland in the desert of southern New Mexico. The location happened to be dubbed *Jornada del Muerto* (Journey of Death) by Spanish explorers centuries earlier. The project was codenamed "Trinity" by the US military and scientists who designed and produced the device being tested. At that exact moment, the first atomic bomb ever built was detonated. This massive test proved beyond any doubt that Einstein's theory was absolutely correct – **matter is composed of energy**.

2.9   The Conversion Process Works Both Ways

Einstein's renowned $E=mc^2$ demonstrates that an object can lose mass as a result of a chemical or physical reaction when some of its mass is converted to energy. As with any mathematical equation, Einstein's formula works both ways. Just as energy can be released by converting some of an object's mass to energy, energy can be absorbed by an object, and thereby increase its mass. This conversion process actually happens at the molecular level. Individual atoms group together to form molecules.

A chemical reaction is the process of rearranging atoms. Chemistry involves the exchange of atoms between molecules and the configuration of new molecules. Existing molecules can be split into smaller molecules or completely broken up into separate atoms. In all of these chemical reactions, energy is released by some atoms and absorbed by others. Energy is released and absorbed by electrons in atoms in the form of photons.

2.10   Chemistry and the Structure of Matter

Scientists have determined that the energy that constitutes matter has a very specific configuration. Based on experimentation, scientists have formulated a systematic theory for understanding how energy in matter is organized and how it behaves. When studying matter and objects, the physics of chemistry come into focus.

Chemistry, as mentioned above, is the study of interactions between atoms to form molecules. Individual atoms are very tiny; a typical atom is only about two hundred millionths of a centimeter in diameter, which is why we cannot see individual atoms. But we do see atoms when they are grouped together, as in any solid object that is clearly visible to the eye.

Scientists have identified 113 different types of atoms, which are called "elements," although only 99 occur in nature, the other

14 were created artificially by scientists using particle accelerators. (Only a few atoms of these man-made elements have ever existed and none survived for more than a few seconds after their creation before decaying radioactively into atomic debris).

All the elements are configured according to the same structure, but they differ in size and number of components. Their individual differences give each element a unique set of characteristics.

2.11   The Electrical Substance of Atoms

Atoms contain positive and negative forces which attract one another. Tiny bundles of negative energy called "electrons" circulate in an atom. Although they are constantly moving, electrons are held to the atom by their attraction to tiny bundles of positive energy called "protons." Protons (located at the center of the atom in an area called the "nucleus") are stationary, especially when compared to electrons which circulate in orbits around the nucleus. Individual atoms have an equal number of positive protons and negative electrons and because they balance each other, an atom is electrically neutral.

There is an inherent paradox in the structure of atoms. Although positive and negative forces are attracted to each other, they are repelled by the same kind of force. Protons repel other protons, and electrons repel other electrons. This characteristic explains why electrons maintain distance between each other by constantly orbiting around the atomic nucleus. But how do stationary protons remain closely packed together in the atomic nucleus? This phenomenon is possible because of a third sub-atomic particle called a "neutron," also located within the atomic nucleus. Neutrons, which are about the same size as protons, are able to shield the protons from each other sufficiently to offset their repulsion of each other.

Another inherent paradox in the structure of the atom has to do with the electrons and protons. Since there is a strong force of attraction between them, why do the electrons circle around the protons and not cling to them? It seems that the collapse of the atomic structure (which would occur if the electrons attached themselves to the protons) is prevented by other fundamental forms of energy also exert force over the sub-atomic particles. In addition to electro-magnetism, the three other forces active within an atom are: "gravity," the "strong force," and the "weak force." These forces exist in opposition to one another and create a tension sufficient to maintain the fundamental configuration of atoms.

A third paradox concerns the size of electrons. Although an electron contains enough negative electrical energy to offset the positive electrical energy of a proton, an electron is much smaller than a proton. It would take about 2,000 electrons to equal the same mass as a single proton. Despite their smaller size, electrons also exert a powerful repulsive force towards each other. In order to maintain distance between electrons, they arrange themselves into orbits at various elevations from the nucleus.

Electrons are able to balance their attraction to protons at the center of the atom and their repulsion to each other. The electrons arrange themselves in a pattern of orbits called "shells." The inner shells surrounding the atom are spherical but the outer shells are more complicated. The closest shell to the nucleus can have up to two orbiting electrons.

The next outermost shell can have up to eighteen electrons orbiting (and so on...). The importance of this configuration, with regards to low lever laser therapy, is that the division between shells creates a lot of empty space within an atom relative to the size of the sub-atomic particles. The atomic nucleus at the center of an atom takes up only one trillionth of the volume of the atom and its diameter is only one hundred thousandth of the atom's diameter.

Most of an atom is comprised of empty space between the electrons, which they maintain so as to distance themselves from each other. This explains why light and other forms of electromagnetic energy are able to simply pass through solid objects without being absorbed. When compared to a photon, the size of an atom is huge. There is plenty of room for the photons to flow through atoms (and avoid collision with any subatomic particles) and continue on their path uninterrupted.

Some photons are absorbed by electrons. The electrons that orbit further away from the atomic nucleus have more energy. There is a reason for this. Before an electron "jumps" to a shell that is further out, it must absorb some energy. Electrons are energized by additional electricity, heat or light, such as that provided by low level laser energy. The amount of energy a particular electron will respond to, is determined by the element type and the orbit level it occupied within that atom. When an electron absorbs the specific amount of energy it needs, it can move out of its present orbit, a circumstance referred to by physicists as a "quantum leap."

2.12 Electrons Release Energy and Cause Atoms to Interact

When an electron releases excess energy, it emits a photon. An electron emits the same amount of energy that it absorbs upon making a quantum leap. A photon emitted by an electron possesses a particular wavelength. Photon wavelengths form the basis for spectroscopy, a scientific diagnostic method which identifies unknown materials by analyzing the wavelengths of light they emit. In a chemistry laboratory, the material being analyzed is heated. In an astronomy observatory, the light and radiation from a star are analyzed. Both processes rely on the science of spectroscopy.

Atoms do not exist in isolation, except those deep in outer space. A solid object, like a human body, is composed of many different atoms. Atoms naturally interact because of their electrical attraction to each other, and electrical attraction is the basis for chemical reactions. Chemical bonding occurs when atoms combine their outermost electrons in a way that is energetically favorable. The shared electrons are identified as "valence electrons." When atoms bond together, they achieve a lower overall energy than individual atoms. In most chemical reactions, the formation of bonds releases energy and the breaking of bonds require the taking in of energy. The sharing of valence electrons is the mechanism for most chemical reactions.

The electrical attraction between some atoms is so great that instead of sharing electrons, a nearby atom can "steal" an outermost electron from another atom. Atoms and molecules that tend to allow their outermost electrons to be "stolen" are known as "free radicals." There are "good" and "bad" free radicals. For instance, good free radicals are used by the body to selectively kill pathological bacteria. Bad free radicals may be responsible for producing painful inflammation in an arthritic joint. Irradiation with 633nm or 830nm of laser light reduces the number of free radicals produced in anomalous blood cells.

Medical research into causes of aging and degenerative disease has shown that free radicals are usually involved with processes that damage cells. Treatment of free radical pathology using optimal nutrition and dietary supplements has become prevalent in holistic medicine.

2.13   The Human Body as a Chemical Factory

People eat for more reasons than just because they are hungry. We may not know precisely what happens to the food we consume, but many people intuitively understand that the body needs a constant supply of nutrients to stay healthy. Food supplies raw materials that constitute the building blocks of our metabolic processes. The cells in our bodies function like tiny chemical reaction factories. However, even the ingestion of nutritionally perfect food will not help if damage to the cell walls prevents the process of osmosis.

Osmosis is the passage of a solvent (food nutrients surrounded by water molecules) from a more concentrated to a less concentrated solution through a semi permeable membrane (cell wall). This tends to equalize concentrations of the two solutions. Low level laser therapy energizes the cell wall to allow more effective osmosis. Each of the laser inventions reenergize cell membranes so that food nutrients can transfer through cellular membranes to nourish cells and better help them thrive.

Supplies of raw materials are replenished daily by ingestion and digestion, and a variety of molecules are constantly circulated in the blood. The cellular fluid (98% of the body is water) and watery environment inside the body provides a rich environment for thousands of metabolic processes to occur every moment. In fact, scientists have discovered the human body is a vast conglomeration of intricate and complex sequences of chemical reactions; each reaction supplies molecules and energy for the next combination in the sequence. Many of these chemical reactions have been identified and studied by biochemists. As a result of this chemistry, a huge amount of energy is constantly being exchanged within the human body.

While these chemical reactions occur, the human body continues to be animated by our consciousness and filled with our bodily energy or life force (known as *qi* in China, *ki* in Japan and *Prana* in India). However, when the chemistry is disrupted, the entire body is affected. It lacks the energy to function as it should. If such a disruption continues, illness in some area of the body is inevitable. If the disruption is prolonged energy dissipates; the cumulative effect is so pervasive the entire body eventually ceases to function. Everything depends on energy.

2.14   Energy: the Base of Reality

Scientific research conducted during the last 300 years has revealed the importance of energy. Not only does everything depend on energy, everything on earth, including human beings, is energy. The very "stuff" of reality, material objects that we assume are solid, such as the chair we sit on are actually manifestations of energy. We don't see objects in terms of energy because of the minute size of photons and atoms. Despite the physical limitations of our eyes and senses, the logical mind can understand that everything is composed of energy. This piece of information is not merely a belief, but rather it describes reality as scientists have confirmed it to be.

Scientists have found the closer and deeper we investigate; the smaller subatomic particles appear to be. We have already described how atoms are composed of subatomic particles call neutrons, protons, and electrons. These subatomic particles, however, are composed of even smaller bits called "quarks," "gluons" and "muons," which are essentially energy. A collection of complex

25

mathematical formulas called "quantum mechanics" has been developed to define the voluminous experimental evidence that has been gathered. The evidence supports one underlying principle: **everything is energy**.

2.15 Understanding the Forms of Energy

We are subjected to many forms of powerful energy in our everyday lives. We experience this energy directly through our five senses. We feel on our skin the heat radiating from a fire. We feel the forward motion when gravity causes our bicycle to accelerate when going downhill. We feel the force of kinetic energy in the swing when hammering a nail into a thick board. We feel the force of gravity upon jumping down from monkey bars on a playground. The sound waves of a mother's voice calling a child in for dinner will be heard from their force of energy.

We react to various forms of energy! Our eyes squint from the bright sunlight after watching a movie for two hours in a dark theater. We may be startled by a small electric spark when striding on a carpet, or we receive a shock when we touch a doorknob or even another person.

Some forms of energy are more minute and complex. Scientists have developed laboratory instruments to observe, test and measure tiny bursts of energy. Physicists construct mathematical formulas that precisely describe how and why particulate energy forms behave as they do.

The subtle principles and formulas that describe small phenomena can be applied to large projects. Based on physicists' laboratory experiments and carefully worked out formulas, engineers are able to predict how miniscule amounts of energy will react when combined into much larger amounts. For example, after testing small segments of transmission cable, electric utility companies invest in the construction of large, high voltage power lines to transport electricity over long distances. From their calculations of the powerful electromagnetic forces, the electrical engineers are confident that the power lines will operate as intended.

This engineering feat is not unlike what our ancestors did when they constructed Stonehenge. Science is often accomplished in this manner. Laboratories are conducive to small experiments where careful measurements can be made with precise instruments. Such small experiments lead to theories and mathematical formulas, which allow scientists to predict how energy will react in different situations. This process has been used for hundreds of years and is known as the "scientific method." All of the new inventions we enjoy using – automobiles, televisions, telephones, air conditioning and computers – evolved from scientific theory, experimentation and application. Our modern methodology continuously leads to new inventions while perfecting existing technology. **Understanding energy continues to be the most important area of scientific inquiry.**

Chapter 3   The Significance of Light and Laser

3.1   The Significance of Light

Everyone knows what light is. When we open our eyes we see, because all forms are defined by light. Objects cannot be recognized in a totally dark room. Light is fundamental to our lives. Most of our knowledge of the world is based on our sight. Simply through observation, early humans could track the cycles of natural events.

The monument Stonehenge, a circle of massive boulders erected by stone-age tribes in Great Britain, is aligned with planetary cycles and other astronomical events those generations of early people observed in the sky. Our ancestors were so intrigued with the sun, the moon, and all the small lights they saw in the night sky, that they devoted their civilization to this grand natural spectacle. Without fully understanding what they were seeing, their confidence in what they observed inspired them to record their knowledge in stone for all time.

Light always comes from some particular source and direction. When we are out of doors in a natural setting, there is sunlight. At night we might have a campfire; on some nights we can rely on the light of the moon. Indoors, there are candles or electric light bulbs. Light stimulates the imagination and curiosity, not only of physicists, but artists as well. The French painter, Georges De La Tour's works

28

always have pensive figures illuminated by one candle, creating atmospheric extremes of light and dark. There is a story about the great Dutch artist, Rembrandt van Rijn. As a child he worked in a tall, dark windmill. Day after day, he would sit and watch the rhythmic light, as the rotating arms continuously flickered past the one small window. Could this early experience have influenced the artist in his creation of masterworks that vibrate with light emanating from the very faces being depicted?

Today we have a scientific explanation for the question, "what is light?" Physicists, who study the fundamental behavior of energy and objects in the physical world, are specialists on the nature of light and its sources. They have developed complex theories which involve meticulous systems of mathematical formulas. Fortunately for those of us who are not physicists, light is so familiar its energy may be understood without mastering all the mathematics.

3.2   Light As a Fundamental Form of Energy

Light energy is the basis for all existence on planet earth. It is no accident that the opening paragraph of the *Bible*, one of the oldest books ever written and the basis for several great religions, describes the formation of the universe as beginning with light.

"In the beginning, God created the heaven and the earth. Now the earth was unformed and void, and darkness was upon the face of the deep; and the spirit of God hovered over the face of the waters. And God said: 'Let there be light.' And there was light. And God saw the light, that it was good."

Although early civilizations lacked an understanding of science as it is known today, they recognized the fundamental importance of light energy and related it to everything known to exist. We still have reverence for the intricacy and complexity of the energy system of which we are a part. The more we understand about light, the more we can directly benefit from it and nurture our own energy.

3.3   Light Stimulating Biological Activity

The sun bathes the earth in its light and heats the surface of the earth and our atmosphere. Without this sunlight, the earth would be a frozen chunk of rock floating in the dark, lifeless vacuum of space. The earth would resemble other planets in our solar system, such as Mercury, Mars, or Venus, which are not positioned, in proximity to the sun so as to receive the life-giving benefits from our star's sunlight. The light itself is a force that makes life on earth possible. As seen from outer space, we have a blue and green planet bursting with life.

To know that light stimulates biological activity, just look at the growing plants. Green plants depend entirely on photosynthesis for their metabolic processes. Once they sprout from seeds, the life force of plants cannot continue without the energy of sunlight. Plants must have particular wavelengths of light. It does not matter if this light is created naturally or artificially with electric light bulbs. With a sufficient amount and type of light, green plants thrive. Without it, they dry up and die.

3.4   Light Stimulating Vitamin D Production

Just like plant life, human cells require light energy. Light is crucial in the production of vitamin D, essential for the chemical processing (metabolism) of calcium and phosphorus. Without it our bones and muscles would suffer from a variety of ailments, such as cramps, rheumatism, rickets and tooth decay. Fortunately, vitamin D is produced by the skin when it is exposed to the ultraviolet wavelengths of light contained in sunlight and in a tanning salon.

While it is true that we can ingest vitamin D in our food (vitamin D is so crucial to good health that it is added to milk), we won't be deficient as long as we are exposed to an adequate amount of life sustaining light from the sun.

3.5   Light Energy to the Rescue

Normally, the body must supply its own energy for metabolic activity. But what happens if the body is having problems functioning, perhaps, because it has sustained a serious injury or is just slowing down from old age? We can supply it with more raw materials (nutrients) for these chemical reactions, but nutrients alone might not be able to speed up our metabolic chemistry. How can we add the energy that our cells need for chemicals to react faster?

After decades of research dedicated to answering this question, I have developed a method of adding energy to the body which stimulates cellular activity. I discovered that low level laser light, tuned to specific frequencies (wavelengths), stimulates metabolic processes in the human body at the cellular level. In other words, **light can stimulate healing**.

As we saw earlier, Popp shows how DNA emits light and, in turn, the special frequency of light helps repair DNA. Before moving into the mechanisms for using light, especially laser light, to stimulate healing in biological systems, we must first understand more about the very nature of light.

3.6   Moving at the Speed of Light

Unlike other forms of energy, light does not need to transmit its energy through a medium. Rather, light is an energy source unto itself. Sound vibrations, for instance, must travel through air or water before they can be heard. In the same way, the energy of momentum must be transferred to an object to give it motion. But light does not need anything to perpetuate its existence. Light from distant stars travels for billions of years through the emptiness of interstellar space before becoming visible in the night sky on earth.

Because **light is pure energy,** it moves extremely fast. The famous seventeenth century scientist, Galileo Galilei, utilizing the help of assistants, attempted to calculate the speed of light by measuring the time it took to exchange light signals between two distant hilltops. All he could determine was that light travels so fast that the event to be timed was faster than the reaction time of the humans conducting the experiment with him. It was impossible for Galileo to measure the speed of light in that way.

In 1849, the French scientist, A.H. Louis Fizeau, built a cleverly designed laboratory device using a mirror and a rotating wheel to measure the speed of light. A short time later, another French scientist, J.B. Leon Foucault, refined Fizeau's experiment to make the calculation more precise. The experiment was repeated in 1883 by the American physicist, Albert Michelson, who continued to refine and improve the mechanism over the next fifty years. Michelson's last experiment, completed shortly after his death in 1931, gave the value for the speed of light at 299,774 kilometers (Km) per second, with a possible error of 11 kilometers. This number was slightly incorrect. The speed of light, designated by scientists as the letter "**c**" (as in Einstein's famous formula for the energy of light, $E=mc^2$), is now known to be 299,792.5 kilometers per second, with an error margin of 0.1 kilometers.

Physicists have discovered that all light energy travels at this same speed. Whether it's light from a reading lamp, a computer screen, a tanning bed, a neon sign, the sun, or a distant star that has been traveling for all eternity through the limitless vacuums of space - all light covers a distance of 299,792.5 kilometers per second. Albert Einstein, the famous Nobel Prize winning physicist, theorized that the speed of light remains constant and is unaffected by the speed of the source from which the light emanates. He also theorized that nothing, neither a material object nor any other form of energy, including soliton waves, could exceed the speed of light.

3.7   Soliton Waves

A single (solitary) wave propagates with little loss of energy and retains its shape and speed after colliding with another such wave. When two of these solitary energy waves meet, they form what is called a "soliton wave," a special type of non-linear light wave that never changes shape as it travels. A soliton wave is actually half a wave, either the crest or the trough. A wave that doesn't wave was once, but no longer, considered to be impossible. The two solitary waves look exactly as they did before colliding and they never generate small waves during collisions with other soliton waves. According to James Oschman, "A soliton can trap an electric charge and carry it along. Once the soliton is formed, this charge transfer does not require further input of energy; thus a sort of super conductor is created." It turns out that a soliton wave is very robust, even in the face of perturbations. It continues undistorted through objects such as fiber optics, water, air, and the human body. Japanese scientists doing fiber optic research found that soliton data can travel the equivalent of 4500 times around the earth without any loss of information.[5] It is believed that soliton waves also have magnetic properties.[6]

Soliton waves have been observed in nature since before the twentieth century. John Scott Russell in 1849 observed this unusual wave along a canal in Scotland. As horses were pulling a boat through canal waters, Russell noticed that a distinct wave was created. The wave continued to roll along the canal without changing shape at a speed of up to 8 or 9 miles an hour. Another more striking example is the often cataclysmic *tsunami* (tidal wave), which is produced by displacement of the sea floor during an underwater earthquake, volcano or displacement of large amounts of water by huge portions of undermined rock breaking loose and crashing into the ocean.[7]

However, in recent years, scientists have discovered solitons in many other contexts; they are involved in processes that extend from the cosmic level to the subatomic level. They are also of central importance in many diverse biological processes. According to Oschman, solitons resemble nerve impulses in the body. "Various therapists have noted that waves resembling solitons appear from time to time in the body, and seem to have beneficial effects, including the release and/or resolution of traumatic memories."[8]

3.8   Soliton Waves Used in Low Level Lasers

By combining two or more laser diodes which are regulated by microprocessors, I have been able to produce controlled solitons, which are then delivered by low level lasers. The nature of the soliton wave in my Q Laser series enables the energy to penetrate deeply into the body without changing or losing its wave form or the information it carries. This unique process (patent pending) enables my lasers to carry electrons throughout the body to restore damaged cells. When the energy has been restored in a cell membrane, osmosis (transference of nutrients) can better occur. Then a chain reaction, beginning with better cellular health, leading to increased tissue health, improved organ health, and finally maximum overall physiological health, takes place.

The soliton wave, produced by the Q Laser series, is responsible for the superior therapeutic success of these lasers. The higher amplitude soliton, created by a meeting of two solitary waves, creates a very low energy output, which harmonizes (resonates) with the human body's energy. Its subtle energy penetrates deeply into all tissues such as ligaments, joints, bones, blood vessels and organs. This energy also carries electrons that re-energize cell membranes damaged by trauma, pollutants, and other forms of stress.

The same effect takes place in the world's oceans and seas when two waves meet. Subtle energy moves through each wave beneath the surface of the sea to create a soliton. It is just such a soliton wave that allows whales and dolphins to communicate across distances as far as 1000 miles. Thus, in my Q series lasers several different visible and invisible (infrared) wavelengths are combined to control the laser's power density by computer circuitry. The result is a very superior set of therapeutic laser light instruments.

3.9   Constant Oscillation

Physicists have discovered that as light energy travels forward, it also exhibits another form of constant motion, "oscillation," an up-and-down wiggling motion that is superimposed on the forward motion. This pattern of motion resembles the shape of the letter "S" lying on its side (Figure 2). We can visualize oscillation by imagining multiple letter S's on their sides all connected together, end-to-end. As the light energy moves forward, it oscillates around the curves of the S's like a super roller coaster car full of screaming passengers hurtling along. It climbs upwards to the peak of the big hill, then over the top and down into the bowl at the bottom of the hill, and then upwards again, to be repeated over and over (Figure 2). Physicists call this roller coaster-like curvaceous movement, a "wave pattern."

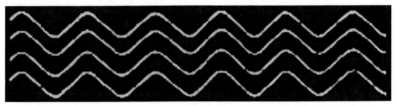

Fig. 2

Because scientists use trigonometry to describe it mathematically, the wave pattern is also referred to as a "sine wave." Light energy that is visible to humans, wiggles through its wave pattern or sine wave approximately three hundred trillion times per second (equivalent to 3×10^{14} in scientific notation). Scientists cite the number of oscillations per second in terms of *hertz*, named after the nineteenth century German scientist, Heinrich Hertz, who conducted experiments on electromagnetic waves. The course of the history of science was changed forever by Hertz's discovery of radio waves.

3.10   Measuring the Wavelength of Visible Light

Although all light travels at the same speed and oscillates up-and-down as it moves forward, physicists have discovered that all light is not the same. A laboratory instrument called an "interferometer" can measure the distance from the peak (or crest) of one wave to the peak of the next wave (Figure 3). A single measurement is known as a "wavelength."

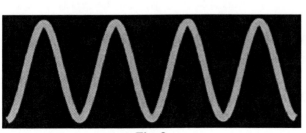

Fig. 3

The wavelengths of visible

light are so tiny they are measured in terms of billionths of a meter, designated as "nanometers" (about forty millionths of an inch). The individual nanometer is also referred to by scientists as an "angstrom."

The human eye, a sensory organ that detects light, is limited in its ability. Most light cannot be observed because its wavelength is too short or too long for the eyes to process. The wavelength of visible light is between 390 and 770 nanometers.

Each wavelength corresponds to a color. Violet is the shortest visible wavelength, and red is the longest. Between violet and red are all the colors of the rainbow: violet (purple), indigo (purplish blue), blue, green, yellow, orange, and red. Any wavelength shorter than violet (ultraviolet), cannot be seen by the human eye. Any wavelength longer than red (infra-red), also cannot be seen by humans.

As mentioned above, all light oscillates in a wavelike motion as it travels. Just as light can have different wavelengths, the number of oscillations per second, referred to as "frequency," also varies. Shorter wavelengths oscillate faster because it takes less time for the pattern to repeat itself from crest to crest. Longer wavelengths oscillate more slowly because the crests are farther apart. Hence, the frequency of oscillation corresponds to the light's wavelength. The frequency of visible light is approximately three hundred trillion times per second.

3.11   The Invisible Energy All around Us

Many forms of energy exhibit the same wavelike behavior as visible light. At the far end of short waves are x-rays, which doctors use to take photographs of bones inside the human body. Longer waves include microwaves, used in microwave ovens and cell phones. Radio waves transmit signals for communication and entertainment to our cellular telephones, radios, and televisions. We can't see any of this energy even though it closely resembles visible light in structure and behavior. The difference in wavelength is enough to make it invisible to us.

All of these various types of energy, whether visible or not, are subtle forms of electromagnetic radiation or electromagnetic energy. **Low level laser light is a healing form of electromagnetic energy destined to have a significant effect on human life**. It will produce many dramatic benefits that are already beginning to manifest themselves. Some are described in this book.

3.12   Laser: Concentrated Light

The word "LASER" is an acronym for Light Amplification by the Stimulated Emission of Radiation. The usual laser device produces a narrow, bright beam of colored light of unusual quality. Unlike the beam of light from an ordinary flashlight that becomes dimmer as it spreads out, a beam of laser light tends to hold its intensity as it travels forward until absorbed by some medium.

The first lasers were tube lasers. Their central core is filled with a material composed of atoms that become the "lasing atoms." One end of the tube-shaped core is completely coated with a silvery film that acts like an ordinary mirror and reflects back all the light that hits it. The other end has holes in the silvery film that reflects some light and lets the rest pass through. The "lasing atoms" in the core material remain in a resting state until the core material is energized by a flash of light or a burst of electricity. Normally, the flash of light stimulates the electrons in the core material to momentarily increase their energy and then to release the excess energy, spontaneously (in a process known as decay) by emitting photons. Moreover, the photons emitted by the electrons' decay are normally emitted randomly in any direction.

But the design of the laser changes this randomized emission process. During the brief moment that an electron in a "lasing atom" is energized, but before it decays, it is hit with another photon of the same wavelength as the one it is about to emit. Then, instead of one photon being emitted, two photons are now moving in synchronization at the same wavelength and in the same direction. Where there was once a single photon, there are now two, exactly the same. These two photons will hit two more electrons and the same thing will happen. But now there are four photons that are exactly alike. The chain reaction continues like an avalanche.

Mirrors in the lasing device enhance the process of the doubling of photons. The back and forth reflection of the photons between the mirrors increases their numbers enormously. This chain reaction is known as "amplification." A steady stream of photons escapes through the partially mirrored end of the core. The released photons are all the same wavelength and are aligned so the peaks and valleys of their oscillating motion correspond. Such wavelength alignment is known as being *"in phase."* Light that possesses all the same wavelength, and is in phase, is called "coherent," and it is coherence that makes laser light so unique. Regular light bulbs or LEDs (light emitting diodes) are not coherent and do not provide the same benefits as coherent laser light.

3.13   Various Uses of Lasers

Albert Einstein worked out the theory of how lasers function in 1917. Then the development of quantum mechanics and the proliferation of mathematical formulas that describe the behavior of subatomic particles further advanced the theory of laser light. However, it was not until May 16, 1960, that the first fully functioning laser device was tested by an American scientist. Theodore Maiman, working at the Hughes Research Laboratories in Malibu, California, presented a demonstration of lasering that would legitimize the technology in the eyes of the scientific world. In the decades that followed, scientists and inventors developed many types of lasers and multiple uses for them. I have put together what I believe are the best low level lasers in the market place.

Lasers do many things. They can create straight beams of visible light, useful for determining elevation and alignment when surveying. The straight beam is also useful for aligning building foundations and steel or wood beam infrastructures. Civil engineers use straight beams for alignment when constructing tunnels and laying large pipes.

The National Aeronautics and Space Administration (NASA) uses lasers for designing satellites and for experiments conducted in outer space. The military incorporates lasers into its new "smart bombs" to guide them to their targets.

Lasers are also incorporated into home entertainment electronic devices. They read the data embedded in the silver disks used for compact disk (CD) music players and digital video disk (DVD) players.

We encounter lasers whenever we shop at a supermarket or grocery store that scans the bar codes printed on product labels at the checkout counter. The red light that is emitted through the small glass window set into the counter, or from the hand-held device used by the checkout clerk, is a laser.

Many computer devices are attached to printers that use lasers to activate a type of photocopier "Xerox" printing. These printers can reproduce images with a precision not possible by other inexpensive printing technologies.

Lasers are also increasingly being used in the medical profession. Surgeons routinely employ lasers in various types of corrective operations. Ophthalmologists, for instance, correct the eyeball with laser surgery. Dermatologists and reconstructive surgeons take advantage of the cutting ability of laser light to accomplish plastic skin surgery.

Lasers are so important in our society that if lasers were suddenly not available our society would come to a standstill.

3.14   The U.S. Military's Use of Laser: "Star Wars?"

On November 5, 2002, the U.S. Army announced that it had successfully tested a prototype defensive weapon over the White Sands test range in New Mexico. A high-energy laser was able to shoot down an artillery shell in mid-flight. The Army and the manufacturer that developed the weapon, TRW Inc., made a joint statement that the laser tracked, locked onto, and fired a burst of concentrated light energy photons at an explosive flying through the air faster than the speed of sound. The Army's Space and Missile Defense Commander went on to say: "Seconds later, at a point well short of its intended destination, the projectile was destroyed."

This new laser weapon, the *Mobile Tactical High Energy Laser* (MTHEL), demonstrates another use for lasers that will change the way future wars will be conducted. "This shoot-down shifts the paradigm for defensive capabilities. We've shown that even an artillery projectile hurtling through the air at supersonic speed is no match for a laser," said Army Lt. Gen. Joseph Cosumano, head of the Missile Defense Command. "Tactical high energy lasers have the capacity to change the face of the battlefield," he added.

3.15   Different Types of Lasers

All lasers have the same basic design and contain certain primary components. They are: (1) an active medium; (2) an excitation mechanism; (3) a high reflective mirror; and (4) a mirror allowing partial transmission. These four components are arranged so that they produce (amplify) a thin, intense beam of light in which all the photons are aligned (coherent). Laser light consists of a single wavelength of light (monochromatic), or a narrow range of wavelengths. Lasers can produce light (ultraviolet and infrared) that is invisible to the human eye.

The active medium in the laser's core, the lasing material, can be made of many different substances which have electrons that can be excited to an elevated energy level by an external energy source. The active medium may be a solid crystal (also known as "solid-state laser material") such as Neodymium doped Yttrium Aluminum Garnet (*Nd:YAG*). Laser light can also be focused on a small glass tube filled with a liquid dye (such as 7-diethylamino-4-methycoumarin), which is tunable across ranges of wavelengths with the use of a diffraction grating, instead of the usual fully-reflecting rear mirror. The lasing material can also be a semiconductor composed of such metals as Gallium-Arsenic (*GaAs*). Solid-state lasers use a crystal, a piece of glass or a semiconductor as the light amplifying substance to produce a pulsating beam.

Lasers that employ gases as the medium generate a continuous beam. The most common type of gas laser has a core made of a mixture of helium and neon gas held in a glass tube. Just like a fluorescent light, the electrons are excited when electricity is passed through the gas. This type of laser produces a beam of red light at a wavelength of 633 nanometers. The beam, equivalent to a few thousandths of a watt or less, is not powerful; one can see the light, but not feel its warmth. Another type of gas laser produces photons that move through argon gas and produce a green light with a wavelength of 514 nanometers. A very powerful laser uses carbon-dioxide gas. The light it emits is invisible because it functions in the infrared range at a wavelength of 10,600 nanometers. A carbon-dioxide laser used for medical purposes generates a fifty-watt light beam which can vaporize skin. An even more powerful version of the carbon dioxide laser, that generates thousands of watts of light, is designed to be used in factories to cut metal.

As mentioned above, a laser requires an excitation mechanism, also known as a "pump," to inject energy into the lasing material. The three basic methods for injecting energy are: optical (using a bright light or another laser); electrical (using a current in a gas laser); or by means of a chemical reaction.

The functioning laser also requires a mirror, which reflects 100% of the laser light back towards the lasing material. This mirror, known as the "active medium," triggers the photon chain reaction inside the core.

At the other end of the core is a mirror that reflects less than 100% of the laser light. It is called "the mirror that allows partial transmission." The light that is not reflected back towards the core by this semi-transparent mirror is allowed to escape as a coherent beam.

3.16 Designed to Produce a Specific Wavelength of Light

Each type of laser is designed to produce a specific wavelength of light. The material in the device's core that provides the lasing atoms determines the wavelength of the photons. A laser that is *tunable* to specific wavelengths is referred to as a "dye laser." The lasing material in the core that contains pigment-like molecules *tuned* to produce the particular wavelength of light, also includes a combination of core filters and lenses known as a "diffraction grating." The wavelengths that are screened out do not participate in the chain reaction that produces the laser light.

Most non-computerized diode lasers cannot control the wavelength and power density precisely enough and cause the body to set up a polarity or impedance to this inconsistent type of energy. To put this another way, the body blocks out the energy and does not allow it to enter the cells. This is why all lasers are not "created equal."

3.17 Why Higher Energy is not Beneficial for all Types of Tissue

The laser industry categorizes laser products that deliver less than 1 watt (1000 milliwatts [mw]) of energy as low level lasers. Higher energy lasers, designed for cutting, are used only by medical and engineering professionals who have the proper safety training. During the last 40 years, throughout the world, low level lasers of varying wavelengths have been used for many different conditions. Unexpected and inconsistent results prompted those in the laser industry to try to solve the problem by increasing the power of the instruments. These higher powered low level lasers have gotten varying results, but in general have been inconsistent due to the protective mechanisms of the human body—polarization or impedance.

Impedance sets in when the body is repeatedly exposed to a higher powered low level laser stimulus. A blockage or polarization takes place. This protective polarization mechanism can also occur when magnets are used to treat health problems in both people and animals. At first the magnet is beneficial but, the benefit decreases over time and in some instances, the magnet actually begins to do harm. What happens in these cases? The body polarizes itself and develops impedance to many of these skin-placed electrical and sound-stimulating devices. However, this does not occur with subtle energy lasers that employ soliton waves, such as our patented Q Series laser devices.

The body enjoys receiving gentle applications of subtle energy furnished with soliton waves. To illustrate, imagine being stuck in a room full of negatively opinionated people. You feel uncomfortable from the negative energy they are giving off and try to leave as soon as possible. Your preference would be to mingle with those who give off positive, powerful and uplifting energy, and so you gravitate to them. The body responds to lasers in much the same way. Properly delivered subtle energy is the answer to many disorders. More is not necessarily better. For organs, glands and bellies of muscles, less is definitely better. The subtlety and safety of low level laser will be discussed in the next chapter.

Chapter 4   What is Low Level Laser Therapy?

4.1   Safety of Low level Lasers

For legal reasons, every electronic device carries a label warning about the danger of possible injury. But are most low level lasers really dangerous? The answer is **No!**

Consider the lasers used by cashiers at food stores to read bar codes printed on product labels. Those lasers are not dangerous for two reasons. First, the power of the scanner bean is minimal and considered a Class I device. The scanner will not damage the eye. Second, the natural act of blinking offers further protection, even if one looked directly into the laser light. However, because laser light can be concentrated into a narrow beam which focuses on a small spot such as in hand-held laser pointers, it is much brighter than conventional light. Shining such a concentrated beam of light into the eye is not recommended.

A quick glance at the sun does not injure our eyes. We do it all the time. But if we stared directly at the sun for a long period of time, we risk damaging the retina. In the same way, low power lasers will not damage the eyes. But if someone deliberately stares at the light for a period of time, it could be dangerous. One should never misuse a laser, even a low level laser, for any reason.

There is another important distinction. The common consumer lasers used in department stores to read bar codes are totally different than the industrial lasers used in factories to fabricate metal. Lasers, in fact, can be powerful enough to cut metal. The cutting edge of such industrial lasers begins with a wide beam that is focused by bending the light waves symmetrically using a short-focal-length lens, which concentrates the light at an intense focal spot. The laser beam's power per unit area (*heat flux*) can be extremely high. The light that pulses from a commercial/industrial cutting laser almost instantly eats into and ablates (melts) the metal at the focal point. If you were to look at this type of laser without safety glasses your eye would be permanently damaged in a fraction of a second.

The low level lasers described in this book are not powerful enough to do any cutting. Regardless of how long they are used, they will not create damage. Low level laser therapy works at the cellular level to restore the cells to health. In fact, low level laser instruments are sometimes referred to as "healing lights!" The United States Food and Drug Administration (FDA) has investigated the safety of laser instruments and, in its usual conservative and low-key manner, officially refer to Class 1 low level lasers as "non-significant risk devices."

4.2   Warnings About Misuse of Laser Pointers

Despite their acknowledgement of the safety of low level lasers, the FDA does warn parents and school officials about the potential danger of hand-held laser pointers in the classroom. The FDA, alarmed about the possibility of eye damage to children from such hand-held laser pointers, wants people to know that these products are generally safe when employed as intended by teachers and lecturers to highlight areas on a chart or screen. But it is not unusual for children to use them inappropriately. Price reductions have led to their wider marketing and the FDA is particularly concerned about the promotion and use of these laser products as children's toys. They warn that laser pointers aimed into the eye can cause more damage than staring directly into the sun. Federal law requires a warning on the product label about this potential hazard.

As explained by FDA Lead Deputy Commissioner, Michael A. Friedman, M.D, "These laser pointers are not toys. Parents should treat them with the appropriate care. They are useful tools for adults that may be adapted by children only with adequate supervision." The FDA's official warning in this manner is prompted by its concern that even momentary exposure from a laser pointer, such as might occur from an inadvertent sweep of the light across a person's eyes, will cause temporary flash blindness. Even temporary blindness is dangerous if the exposed person is engaged in a vision-critical activity such as driving a car.

As the FDA points out, it is not the laser light that is dangerous; it is the misuse of the device by a foolish or mischievous child that could be harmful. When laser lights are employed for the purposes for which they are designed, they carry no hazard. As stated, the FDA has issued safety guidelines for manufacturers and operators of laser light shows, and the agency officials' state publicly: "**When used safely, low level lasers are not dangerous.**"

4.3   Laser Safety Categories

The intensity of the light beams produced by lasers depends on the design and the pump that injects energy into the core. The FDA classifies lasers into four categories based on the intensity of laser light output and the potential risk to the eye. Class 1 represents the least risk, and as the power gets progressively higher in Class 2, Class 3A and 3B, and Class 4 lasers, the risk for eye damage increases. The following is a summary of laser classification taken from ANSI Z136.1-1993, section 3 and other sources.[9, 10, 11]

A Class 1 laser, described by the FDA as a "non-significant risk device," is considered safe based upon current medical knowledge. This Class 1 category includes all lasers or laser systems, which cannot emit levels of optical radiation above exposure limits, tolerated by the eye under any exposure conditions inherent in the design of the laser product. Class 1 can include a laser more hazardous than Class 1 parameters allow, if it is embedded in an enclosure that does not allow harmful amounts of radiation to escape.

A **Class 2 laser** or laser system must emit a visible laser beam. Because of its brightness, Class 2 laser light will be too dazzling to stare into for extended periods. Momentary viewing is not considered hazardous since the upper radiant power limit on this type of device is less than the Maximum Permissible Exposure (MPE) for a momentary exposure of 0.25 seconds or less. Intentional extended viewing of a Class 2 laser is considered hazardous to the eye. No known skin exposure hazard or fire hazard exists. Lasers used to read bar codes in stores belong in this category.

Class 3 lasers are divided into two categories, **Class 3a** and **Class 3b**. Class 3a includes most laser pointers on the market today and emits a visible blue, green or red light. These lasers are considered more hazardous to the eyes than a Class 1 or Class 2 laser because they are focused and should not be looked at directly at close range for extended periods of time. Because the eye blinks when it is exposed to a bright light, a temporary exposure to the eye will not cause permanent eye damage. A Class 3b laser is usually an infrared laser. That is, it is not within the visible spectrum of the eye and cannot be seen; therefore the blinking effect is not activated and this class of laser can cause damage to the eye if misused. DO NOT look directly into this laser unless wearing protective eye glasses. Class 3a and Class 3b lasers are not fire hazards, nor do they constitute a hazard to the skin. Any continuous wave (CW) laser that is not Class 1 or Class 2 is a Class 3 device if its output power is one-half watt or less. Class 3a and 3b low level lasers are required to have a Danger Label and Output Aperture Label attached to the laser and/or equipment.

A **Class 4 laser** or laser system is any device that exceeds the output limits, Accessible Emission Limits (AEL's) of a Class 3 device. As would be expected, these lasers may be a visual hazard and protective glasses are required. Cutting lasers belong to this classification. Very stringent control measures are mandated by the FDA for use of Class 4 lasers.

4.4.   Safety of Q Series Lasers.

The Resonator and Rotary Multiplex are predecessors of the Q100 and Q1000 lasers. These lasers have been studied by Dr Sanford R. Simon, Professor of Biochemistry and Pathology, Stony Brook Health Sciences Center, State University of New York. Dr Simon evaluated the lasers on various primary human cell cultures and the study concluded that the Rotary Multiplex, now called the Q1000, was non toxic to all cells. These lasers have also been studied by Underwriters Lab, and based on those studies they have been given the FDA classification of **Class 1**. This means when these instruments are used for treatment, the low level laser light emitted has no potential for harming the user or the recipient. This class of low level laser device is what the governmental agencies indicate is "non-significant risk (NSR)." Unquestionably, low level laser therapy is completely safe.

The 660 Enhancer is classified as a Class **3a** device. The same precautions for the laser pointer should be taken with it–that is don't point it in anyone's eye. The 808 Enhancer is classified as a Class **3b** device and can be harmful to the eyes if looked at directly for any length of time. Since the 808 Enhancer produces invisible infrared light, it should not be used around the eyes. Precautions should be taken so that the eyes are never accidentally exposed to the light. That makes the 808 Enhancer potentially more dangerous than the 660 Enhancer.

4.5   Laser Medicine, World Medicine

Lasers are being used throughout the world. Medical specialists can join the American Society for Laser Medicine and Surgery, the Academy of Laser Dentistry, the International Society for Lasers in Dentistry, the North American Association for Laser Therapy and the World Association for Laser Therapy.

The FDA in 1996 was instructed by Congress to study low level laser devices and their efficacy for treating disease. The FDA as well as the NIH (National Institutes of Health), through its CAM (Complementary and Alternative Medicine) division, is in the process of carrying out several such studies on the efficacy of low level laser therapy. Since the FDA (at this time) requires the study of the effect of low level laser therapy on each disease (and there are hundreds of classifications of diseases), it will take generations for them to complete the studies to show the many benefits of low level laser therapy. As of 2004, the FDA has approved low level lasers for myofascial pain of the shoulder and carpal tunnel syndrome. It is hoped that the FDA will soon change their protocol and conduct general studies of the effect of low level lasers at the cellular level ---- where they actually do their work. Since the body is composed of cells it makes sense to study the affect of low level laser therapy on cells rather than on the many diseases.

The FDA also has already approved many higher powered cutting lasers for specific therapeutic purposes. Many types of surgeons now use lasers to vaporize cellular tissue instead of cutting it the old-fashioned way with metal scalpels. Surgeons have discovered that they can be more accurate with lasers than with scalpels. In effect, medical lasers provide the surgeon with a very precise scalpel. This is especially important in eye surgery where cutting too much would severely harm the patient. Also, the heat produced by the laser can sterilize and cauterize surrounding tissue reducing infection and bleeding.

Dermatologists and plastic surgeons use lasers for treating skin problems. Lasers can remove tattoos and natural skin blemishes. They also remove unwanted hair on the face or other parts of the body.

The FDA has approved lasers for eye surgery. Ophthalmologists make use of lasers to perform two procedures to correct nearsightedness and astigmatism in adults. The Photo Refractive Keratectomy (PRK) is a procedure where the surgeon uses an Excimer laser to remove surface tissue on the cornea. After the eye is numbed with a topical eye drop anesthesia, the patient is told to stare at a fixation light. The surgeon then uses the Excimer laser to remove superficial layers and reshape the affected cornea. Depending on the level of myopia, the thickness of the removed tissue may range from 10 up to 150 microns. By comparison, the typical thickness of a human hair is 125 microns. The laser process lasts under a minute. After the eyes heal, the patient's vision is normal and they need not wear eyeglasses.

A newer procedure, called LASIK (Laser in-Situ Keratomileusis), combines the technology of an Excimer laser with a cutting device called a *microkeratome* to change the shape of the corneal surface. The surgeon uses the microkeratome to precisely cut a flap in the outer layers of the cornea. A small amount of the targeted tissue beneath it is removed with the laser, and then the flap is replaced. The benefit of the microkeratome is that it keeps the surface layer intact, so that only the edge needs to heal. This procedure also eliminates the patient's need for eyeglasses.

Dentists have used low level lasers for more than a decade to treat soft tissue and gum problems. More recently the FDA has approved higher powered cutting lasers for preparing teeth for fillings and crowns, as well as for cutting bone and sterilizing root canals. Cardiologists use lasers to perform vascular surgery. General surgeons remove kidney stones with laser light.

Given the fact that the FDA has already approved many medical laser procedures, it seems likely that the use of low level devices for the treatment of nearly every type of dysfunction will gain approval in the near future. Such approval will allow any person to use this safe and inexpensive low level laser therapy for the relief of his or her osteoarthritis. Since there is very little risk to the self-treating arthritic person, the FDA will probably approve low level lasers for home use by all Americans.

4.6 Positive Evaluations of Low Level Laser Therapy

The FDA has determined that exposure to low level laser light is safe for human beings. The FDA's experience with other laser devices approved for use by doctors has inclined them to have an "open mind" towards low level lasers. Recognizing that the risk to a patient's health posed by low level lasers is very little to none at all, the FDA has shown uncommon restraint in its oversight authority.

When applying for approval, the laser manufacturer can file in two ways; one is to do the necessary clinical studies through an Investigational Review Board (IRB) and apply for a Pre-Market Notification (PMN) stating that the company intends to introduce a device into commercial distribution. The only medical claims the manufacturer is allowed to make are those included in the product description, which is part of the Pre-Market Notification. The company may also apply to the FDA under regulation 510(k). Those making this application, based on existing approvals already granted by the FDA, must have the same type of lasers, but the devices may have an improved delivery system.

The FDA has already issued approvals for low level laser therapy devices and does not appear to have any objection to issuing more in the future. Some companies have made such advances in the laser therapy field that they've recently won FDA approval. Other applications requiring official approval are still pending. In fact the U.S. Government is using low level lasers in its own health care facilities, such as the Veterans' Administration hospitals, and the Military is doing research on LLLT.

Russia is using low level energy in its Cosmonaut program with the aim of enhancing the uptake of nutrients. Scientists have learned that space travel, because it is beyond the field of gravity, affects the way nutrients pass through the cell wall. Basic science teaches that osmosis will not occur through a depolarized membrane and apparently space travel depolarizes or causes the loss of electrons in the cell membranes. In their attempt to solve this problem, the Russians have developed the Scanar unit. This is not a laser but an energy device that uses a complicated method of changing the wave form, the frequency and the power density, to "trick" the body into allowing energy into the cells.

By using advanced computer programming and the soliton wave, the Q Series lasers have solved the problem of polarization. Therefore, the energy gets to the cells and restores the lost electrons.

4.7   Low Level Laser Light Permeating the Body

Low level lasers, including those used to read bar codes, and laser pointers, are not powerful enough to cut or vaporize cell tissue, as do the lasers used by doctors. Low level lasers work through **biostimulation and photostimulation.**

Because light is pure energy, photons penetrate through the skin and the underlying structures. If the light is programmed to the proper wavelength, after penetrating the body it interacts with the atoms of the muscles, organs and other tissues. The photons energize the atoms causing chemical reactions to occur.

The practical, therapeutic effect of laser light was illustrated in two clinical studies carried out on children with cancer. In one study, four Russian oncologists administered Low Level Laser Therapy during the year 2000 to pediatric patients suffering from malignant neoplasm. The children were experiencing complications associated with chemo and radiation therapy. The low intensity laser light alleviated the complications. In another study, the same four oncologists used low level laser therapy to increase mononuclear levels of donors' blood, which in turn led to the physiological release of major factors of immune response development.[12, 13]

4.8   Low Level Laser Light and Biostimulation –
An Alternative to Drugs

Laser **Biostimulation** means that light energy is used to enhance the body's inherent capacity to heal itself. This type of healing differs from conventional medicine commonly practiced by allopathic physicians in our society today.

Organized medicine emphasizes cause and effect. If the cause cannot be found, which is often the case in chronic degenerative diseases, then suppress the symptoms with drugs. If drugs don't work, then do surgery. If surgery doesn't work, send the patient to a counselor who will help them learn to live with the problem. America is getting fed up with this costly and inefficient medicine and is turning to alternative medicine for answers. Prescription drugs developed by pharmaceutical companies target a particular part of the body and trigger a response that hopefully offsets the troublesome ailment by eliciting some sort of quick reaction which reduces the patient's symptoms and discomfort. Many drugs actually suppress the immune system rather than stimulate it. But this is changing. In 2004, there were more visits to complementary, alternative type practitioners than appointments with regular allopathic doctors.

At the fifth Complementary and Alternative Medicine Conference in St Louis, MO, in 2004, Dr. Simon Yu, a member of the American Academy of Family Physicians, read the following excerpts from *The Journal of the American Medical Association*, July 26, 2000.

Medical treatments for disease often focus on treating the symptoms without addressing the underlying causes of the problems. In fact conventional treatments have become the third leading cause of death, with 225,000 deaths after heart disease and cancer." And another quote from the respected JAMA (Nov.11, 1998, 280, 1569-1575) states: "The public has been looking for safer alternative medical care. More than 630 million visits are made annually to alternative medical practitioners, exceeding the total visits to all U.S. primary care physicians. Americans annually spend more than $60 billion on alternative medical care and the majority of that is paid out of the pocket. The proof is in the numbers above. If conventional medicine was successful in treating illness and chronic conditions, the use of alternative medicine probably wouldn't be increasing so rapidly. Often, alternative medicine is sought out because people are suffering from the syndrome called, 'My doctor said everything is fine, then why do I feel so bad?' These individuals are often abandoned as incurable when the primary measurement of success, the treating and relieving of symptoms, is unsuccessful. Longer lasting success only comes from treating causes of illness, not just the symptoms. [14]

This is good! People are taking responsibility for their own health. This is the way it should be. Although prescription drugs might help the most pressing problems and symptoms, they also produce additional health problems and trigger unwanted side effects. These range from cramps, diarrhea, drowsiness and headache to impotence, liver problems, and so forth. That is why all medications are required by law to contain warnings and instructions to discontinue use if severe side effects occur. It is amusing to watch and listen to a TV ad for drugs. Most of the time it takes longer to list all the risks and side effects than it does to tell about its benefits.

Surgery is also fraught with hazards. Following surgery, the patient's body needs to repair the damage left by the surgical wound. All too often the surgery doesn't help. In some cases surgery even makes things worse. Surgery also carries a risk of severe infection despite the care and precautions of the surgical staff. If an infection develops, the surgical wound will not heal. Hospital infections are serious and more and more antibiotics are less and less effective. A systemic infection can be more lethal than the original problem that the surgery was intended to correct.

4.9   Biostimulation: A Different Type of Medicine

Biostimulation is similar to taking a daily dose of vitamins, minerals, enzymes and other types of dietary supplements; it enhances the human body's inherent mechanism to repair itself, fend off harmful organisms, and maintain healthy metabolism. The immune response is directly enhanced by low level laser therapy which is biostimulation in its purest form. Low level laser therapy restores lost electrons to the cell wall and enhances the osmosis of dietary nutrients. Biostimulation also improves absorption in the gut and helps injured cells assimilate nutrients from the blood stream more efficiently. The laser does not heal; rather, the light stimulates the body to heal itself.

Medical biostimulation is not new. Electrical biostimulation has been used during much of the twentieth century. Two forms, Galvanic Therapy and Transcutaneous Electrical Nerve Stimulation (TENS), have been administered by doctors or offered to patients for home use for many years. Both treatments employ skin pads. When positioned correctly, the pads cause small pulses of electricity to pass into the patient's body and stimulate the muscles. Galvanic Therapy helps establish normal muscle function by stimulating blood circulation to the muscle. The TENS therapy is reported to stimulate the body's inherent neural pain relief mechanisms.

4.10   How Far Does Laser Light Penetrate into the Body?

There is no exact limit to the penetration of laser light. The light becomes weaker as it penetrates deeper. The depth at which the light intensity is so low that no healing or biological effect can result is called the "greatest active depth." This depth is contingent on tissue type and is affected by skin pigmentation and surface conditions such as dark clothing, dark hair or even the presence of dirt. If there is variation in the density and frequency of the light, the body will eventually set up impedance or polarization.

Low level laser light not only effectively penetrates muscle tissue; it also penetrates the more transparent fat tissue. It is of interest to note that laser light can even penetrate bone, which may explain some of the dramatic results observed by veterinarians who treat large animals such as injured racehorses. It is excellent therapy for human bones too, as suggested in the following testimony.

> Last summer I developed a bone spur on the heel of my right foot that resulted in tremendous pain" states Mrs. Adele Polakoff, a homemaker from Hackensack, New Jersey. "Walking was a chore for me, and jogging (my favorite activity) was out of the question. Staff doctors in The Foot Center, a podiatry clinic in Hackensack, wanted me to undergo a series of cortisone shots combined with hot water and electrical stimulation. When this treatment barely improved my condition, they had me wear an orthotic instep device in my shoe. When I questioned them about the prognosis, they responded, 'if this orthotic doesn't work, we'll consider you to be a candidate for an operation on the affected heel bone.'
>
> Well, their shoe device failed to work, and I did NOT want to undergo foot surgery. At this point, and looking for

help everywhere, I called my longtime friend and dentist, Dr. Herbert Yolin of Brookline, Massachusetts, to see if he might have any suggestions. He recommended using low level laser light on a series of acupuncture points and also directly on the sore heel bone," Mrs. Polakoff explains. "The result was quite gratifying. As a result of using my laser instrument, which I had purchased some months before to treat a dental problem, I am once again back to doing all the things I love–jogging with my dog Sasha, going for long walks and dancing!

4.11   The Paradox of Biostimulation

A strange paradox has been observed by users of low level laser therapy for biostimulation, and others involved in energy medicine. Completely opposite results may occur, depending on whether the cells being stimulated by the laser's light are located within a laboratory test-tube (*in vitro*) or within a living creature (*in vivo*). Presumably, any cell can be stimulated by the proper frequency of laser light. But what if it's a cell that you do not want to be stimulated, such as a cancer cell?

Laboratory experiments on the effects of low level laser light on laboratory-grown (*in vitro*) cancer cells reveal that cancer cells can be stimulated by laser light. However, cancer cells located in living creatures, observed during *in vivo* experiments, respond in the exact opposite manner. The *in vivo* experiments on rats have shown that small tumors treated with low level laser therapy can recede and completely disappear. (Laser treatment had no effect on tumors over a certain size.) Scientists theorize that it might be the immune system, rather than the tumor, which becomes stimulated by the laser light. That would account both for the above-mentioned paradox and for the successful cancer treatment results.[15] An excellent rule to follow is, "Never stimulate a cancer cell – Resonate it." Resonating lasers, such as the Q1000, operate at fewer than 5 milliWatts of power while stimulating lasers, such as the 808 Enhancer, operate at 300 milliWatts. Resonating lasers that contain the soliton wave, restore electrons to damaged cells such as cancer cells.

The experimenters noted with satisfaction that despite the difference in the *in vitro* and *in vivo* responses to therapy with low level laser light, both indicate that low level laser therapy provides the cells with biostimulation.

4.12   The Paradox of Over-stimulation

For some unknown reason, most people automatically assume that if something is good, more is better. For example, if drinking a small amount of an alcoholic beverage makes people feel relaxed and jovial, they think that drinking more will make them feel even better. Many of us have experienced first hand that this is NOT true when we wake up the next morning with a throbbing headache. Human beings must be aware of the pitfalls of the logic which causes us to overindulge in a good thing. In medicine too, a little bit of cure is usually better than a lot. Homeopathy, where the more diluted a substance is, the more energetically powerful it gets, is a case in point.

This homeopathic-like effect has proven to be true with low level laser therapy as well. Low level laser therapy biostimulation is very effective. The paradox of low level laser therapy is that over stimulation, the use of more laser light than necessary, can negate the benefits already gained. In science this is called the bell curve (Figure 4). Benefits increase on the incline and are at the maximum at the peak of the curve. They then decrease on the declining side, and recede to the original state.

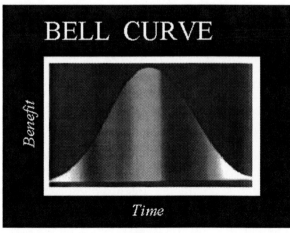

Fig. 4

Although, it does not harm us, as occurs in overmedicating with prescription drugs, using too much low level laser light tends to eliminate benefits that the patient would have received from the proper dose. We must be careful not to use too much laser light in the false belief that excessive biostimulation is beneficial. Both medical practitioners and laypeople use different forms of Kinesiology to determine the correct dosage of low level energy. If there is pain involved, use improvement in the pain level as a method of determining the low level laser dosage. If still in doubt about how long to use the Q Series lasers, refer to the *Low Level Laser User's Manual.* (To order the book call 605-342-5669)

4.13   Not All Low Level Lasers Are the Same

To be effective, lasers designed for low level laser therapy must generate light at wavelengths that are known to cause biostimulation. This is supported by research done in Russia on fruit flies where scientists found that low level laser therapy can have different effects depending on the amount and manner in which it is applied.[16]

There is another important point to be deduced from the on-going low level laser therapy research. Negative results, or results that are not statistically significant, might simply mean that the choice of wavelength or duration of treatment were not optimal for the condition tested. I have, through research and testing, discovered appropriate ways of treating various conditions. These are presented in my *Low Level Laser User's Manual.*[17]

I do foresee the discovery of new forms of low level laser therapy. By incorporating a tiny circuitry control mechanism into my laser design, the Q1000 laser is re-programmable. This means that as more wavelengths are discovered to be effective for various conditions, the laser can be programmed to generate them. From an economic standpoint, each of my laser devices is a good investment; they can easily be repaired and reprogrammed to meet changing needs as new information becomes available in energy medicine. The practice of energy medicine among enlightened and progressive health care professionals is growing in popularity and becoming an important part of Complementary and Alternative Medicine (CAM).

Chapter 5 The Health Benefits of Low Level Laser Therapy

Previously, we talked about low level laser medicine in general. Here, we will discuss the specific health benefits of low level laser therapy.

5.1 The Benefits of Biostimulating Light Waves

The biostimulating light waves emitted by low level lasers increase blood circulation to injured areas, reduce the incidence of infection and infectious organisms, defuse inflammation, relieve pain, stimulate cellular activity and growth in both soft tissue cells, such as those of the skin, and hard tissue cells, such as bone cells.

These advantages combine to promote healing of open wounds, such as diabetic ulcers, and internal inflammatory conditions, such as gastric ulcers. Laser probes stimulate acupuncture points in a fast and simple way; moreover, patients can safely treat themselves and avoid expensive doctor's visits and acupuncture needles.

5.2 The Evidence: Basic Benefits of Low Level Laser Therapy

Physicians are now using light energy to treat many stubborn medical problems. For example, medical doctors at a Veterans'

Affairs hospital recently reported the results of an experiment that combined low level laser light and mild electric stimulation to acupuncture points for Carpal Tunnel Syndrome. The patients' pain, which is characterized by tingling, numbness or pain in the thumb, index (pointer) finger and middle finger, was significantly reduced.[18]

Below is a list compiled from published international reports describing the successful use of low level laser therapy at various hospitals, private medical offices, university medical schools, and in outpatient settings:

- Direct application of low level laser therapy to the skin is highly effective for bed sores, burns, and ulcerations.[19]
- Low level laser therapy influences the proliferation of collagen, elastic fibers, and reticular fibers (fibroblasts), and helps in the regeneration of damaged or dysfunctional lymphatic, muscular, and cartilage tissues.[20]
- Low level laser therapy induces faster healing of post-operative wounds following surgery.[21]
- Low level laser therapy, when administered to the appropriate acupuncture points and directly to trauma sites, is highly effective in closing serious wounds such as those suffered during automobile accidents.[22]
- Low level laser therapy prevents inflammation of the mucous membranes which often results from cancer radiation therapy.[23]
- Low level laser therapy is quite useful for the reduction of pain and general pain management.[24]
- Low level laser therapy counteracts the symptoms of neurological disabilities and neuropathy (numbness, burning, tingling, and itching disorders).[25]
- Low level laser therapy brings about symptomatic relief from degenerative difficulties, such as peripheral nerve disease and spinal cord injuries.[26]
- Low level laser therapy reduces the pain and inflammation associated with sports injuries and other traumas.[27]
- Low level laser therapy diminishes the redness, swelling,

pain, and other inflammatory symptoms of arthritis, including osteoarthritis, rheumatoid arthritis, and gout.[28]

- Low level laser therapy successfully treats diseases of the ear (including the inner ear), nose, and throat.[29, 30]
- Low level laser therapy has been useful in diseases specifically affecting men, such as bacterial chronic prostatitis, benign prostatic hyperplasia, prostate cancer, induratio penis plastica, and urethral stenosis.[31]
- Low level laser therapy has been effectively applied to pediatric problems and gynecological difficulties.[32, 33]
- Low level laser therapy has proven useful for gastrointestinal diseases, including malignant tumors of the gastrointestinal system.[34, 35]
- Low level laser therapy is frequently used for many types of dental problems (See chapters 7 & 8).[36]
- Low level laser therapy effectively stimulates acupuncture points.[37]

The example of Carpal Tunnel Syndrome showcases the clinical effectiveness of low level laser therapy; this is especially important because surgery is often the only effective prescription for this very painful condition. The FDA has approved lasers specifically for the relief of Carpal Tunnel Syndrome.

5.3   Clinical Effectiveness of Low Level Laser Therapy on Carpal Tunnel Syndrome

Symptoms of Carpal Tunnel Syndrome

People suffering from Carpal Tunnel Syndrome often experience numbness while doing ordinary activities such as writing, driving, sewing, holding a book or newspaper and typing. Sometimes the numbness is accompanied by debilitating pain, especially when they wake up in the morning. Other sufferers find themselves dropping things, such as small objects and gadgets that they had been attempting to hold and utilize; the gripping power of a hand affected by Carpal Tunnel Syndrome is weakened or lost altogether over time. The thenar muscles in the hand frequently become so weak that it is impossible to bring the thumb into opposition with the other fingers making it almost impossible to grasp objects. Many patients experience what some may describe as "electrical shocks" emanating from the center of their wrists into their hands.

Progressive Carpal Tunnel Syndrome can become exceedingly painful. Hand discomfort may radiate up the arm to the shoulder and all the way to the neck. Increasing numbness may result in a "blind" hand with loss of feeling in the thumb, index and middle fingers making it almost impossible to do things by feel. Imagine being unable to perform ordinary tasks like buttoning a blouse or fastening a nut onto a bolt where you can't see it. The ability to screw on bolts or do other blind manipulation tasks constitutes the routine work of auto mechanics, cooks, construction workers, tailors, and artists; such incapacitation often forces people out of their occupations.

<u>Causes of Carpal Tunnel Syndrome</u>

Most fundamentally, Carpal Tunnel Syndrome occurs from a pinched nerve in the wrist. Researchers believe that repetitive use of the hand, referred to by health professionals as "repetitive" or "cumulative" trauma disorder, is a prime source of Carpal Tunnel Syndrome. Among both sexes, those who perform word processing on computers, play golf, lift weights, ride bicycles, or work as carpenters hammering boards, lumberjacks chopping trees, or butchers cutting meat, are at risk to develop Carpal Tunnel Syndrome because of the jarring and repetitive activity. Carpal Tunnel Syndrome is a very real occupational hazard.

Interestingly, the majority of Carpal Tunnel Syndrome patients are women between the ages of 40 and 45 years old. Some of the younger female victims experience Carpal Tunnel Syndrome symptoms only while pregnant. A broken wrist, or dislocation of the hand bones, can also harm the median nerve, as can medical conditions involving arthritis, and thyroid problems. Sometimes Carpal Tunnel Syndrome appears for no good reason. One thing is certain— Carpal Tunnel Syndrome almost always gets worse if left untreated.

<u>Anatomy of Carpal Tunnel Syndrome</u>

Carpal Tunnel Syndrome occurs when there is pressure, with subsequent inflammation, on the median nerve where it passes through the wrist in an area known as the carpal tunnel, a narrow passageway near the base of the palm of the hand. Through this opening, the median nerve and the upper extremity's flexor tendons run into the hand. Bones, tendons, and the thick, strong transverse carpal ligament surround the median nerve, which runs into the hand to supply sensation to the thumb, index finger, long middle finger, and half of the ring finger.

The upper extremity of this major nerve also supplies a branch to the muscles of the thumb (the thenar muscles), which move the thumb and are very important for thumb mobility; they enable a person to touch each of the other fingers on the same hand, a motion called *opposition*. The anatomical development of an opposing thumb was crucial in human evolution.

<u>Current Solutions for Carpal Tunnel Syndrome</u>

An informed medical doctor, osteopath, chiropractor, naturopath or other type of health professional might suggest several types of remedies for Carpal Tunnel Syndrome. He or she, for instance, could recommend that the patient avoid tasks that require bending the fingers while flexing the wrist. A Carpal Tunnel Syndrome patient is often told to conscientiously keep the wrist straight; the patient can wear a splint (wrist support) during the day and a brace at night while sleeping to keep the wrist from bending backwards or forwards. Keeping the wrist straight gives the sensitive nerve inside the carpal tunnel more room. There are some wrist exercises that help control Carpal Tunnel Syndrome symptoms. And most importantly, the patient must cease the repetitive motion that aggravates the median nerve. As mentioned above, this may mean changing jobs or giving up a favorite pastime, sport or other activity. It's distressing!

A physician could also prescribe an anti-inflammatory medication to reduce swelling. Cortisone, a steroid hormone, can be injected directly into the carpal tunnel to bring about a reduction in swelling and pain. But if such conventional treatments do not relieve the patient's suffering and disability, the doctor will probably recommend surgery. Each year nearly ½ million Americans have surgeries for Carpal Tunnel Syndrome at the cost of $8,000 to $10,000 per patient, according to the American College of Orthopedic Surgeons.

In addition to low level laser therapy, other complementary modalities for treating Carpal Tunnel Syndrome include vitamin B and thyroid therapy.

The Use of Low Level Laser for Carpal Tunnel Syndrome

Treatment success has increased markedly in recent years because of low level laser therapy. In 1995, rheumatologist, Elhu Wong, M.D., and his colleagues at the University of California, San Francisco, observed that Carpal Tunnel Syndrome patients often had poor posture. Dr. Wong's medical team checked the patients' spinal columns for pain and tenderness by palpating the fifth cervical vertebra (C-5) to the first thoracic vertebra (T-1), and the medial angle of the scapula (shoulder bone). The physicians focused treatment primarily on the posterior neck area, rather than the affected wrists and hands of the thirty-five Carpal Tunnel Syndrome patients in the study. The remedial procedure, strictly consisting of administering low level laser therapy to the tips of the vertebrae, rapidly alleviated pain and tingling in the patients' arms, hands, and fingers.[38] Thirty-three of the thirty-five patients were completely relieved of pain, while the other two had a reduction in their pain.

In another clinical study, which took place two years later, Chicago-based neurologist Martin I. Weintraub, M.D., investigated whether repeated laser light exposure directed to the affected wrist's median nerve could reverse the symptoms and electrophysiological latencies in Carpal Tunnel Syndrome. Thirty patients had consulted Dr. Weintraub for moderate to severe symptoms of Carpal Tunnel Syndrome. The only procedure he followed was application of low level laser therapy at thirty-three-second intervals to five points along the median nerve. They felt no discomfort or heat emission from the laser therapy.

Dr. Weintraub reported in his article that complete resolution of the patients' pretreatment symptoms and/or abnormal physical findings were achieved in seventy-seven percent of the Carpal Tunnel Syndrome cases. Nocturnal complaints were the earliest symptoms to disappear—his patients were able to sleep throughout the night once again. Then the tingling, stiffness and weakness disappeared. Immediately after undergoing their course of low level laser therapy, eleven of the patients resumed their work activities in full.[39]

The FDA, in 2002, approved the use of laser treatment for Carpal Tunnel Syndrome based on a study conducted at the General Motors Corporation. In this double blind, placebo-controlled study, about seventy percent of the injured workers, who had failed to find relief from other treatments, responded to the low level laser therapy treatment. General Motors now offers laser therapy to injured workers in its in-plant medical facilities.

Low level laser therapy had been recognized as therapeutic for the complications of Carpal Tunnel Syndrome even before Drs. Wong and Weintraub had undertaken their individual clinical studies. For example, in 1993 industrial injury specialist, Ishiban Yu, M.D., of Detroit, Michigan utilized low level laser therapy for auto assembly-line workers who had become disabled as a result of repetitive actions involving the tightening of nuts and bolts. Dr. Yu now advises fellow practitioners of industrial medicine to apply low level laser therapy in such cases because of his good to excellent results.[40]

There has been continuous progress in clinical studies of the application of low level laser therapy for Carpal Tunnel Syndrome. Full disclosures of laser treatment effects were presented by Mathew A. Naeser, M.D., of Los Angeles to the 2nd World Congress of the Association for Laser Therapy in Kansas City, Missouri in September 1998.[41]

Fortunately, low level laser therapy can be applied as a safe and effective treatment for many types of injuries and illness-related problems. Moreover, the cost of low level laser therapy treatments usually runs about $25-50 per treatment for a typical course of ten to fifteen treatments. The savings on medical costs from low level laser are huge!

In addition, there are also positive studies in the field of dentistry, which will be discussed in a later chapter. The following cases show the dramatic affects of low level laser therapy on wounds caused by accidents and surgical procedures. The evidence from these cases will no doubt change the minds of skeptics who doubt the effectiveness of low level laser therapy

5.4   Five Examples: Low Level Laser Light to the Rescue

It's Good "Horse Sense" to Use Light for Healing

Light is fundamental to our existence, but can it promote healing in a living creature? A series of photographs that graphically show what happened to a lucky horse that received low level light therapy, answers that question.

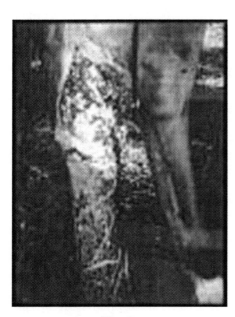

Fig. 5

The first photograph (Figure 5) shows a severe injury to the horse's leg from a barbed wire cut. The sharp wire prongs left a huge gaping wound. For six weeks a Veterinarian tried to heal the animal's leg with any and all techniques commonly used in the veterinary profession, but nothing worked. When the wound didn't heal, infection set in, and was followed by the onset of gangrene. Finally, discouraged by the lack of success, the Vet said that the most humane thing to do was to "put the mare down." Euthanasia (what some people call "mercy killing") is still considered the only option for a horse that loses the use of a leg.

But that advice just didn't make good horse sense to the owner, a South Dakota horse breeder. Instead she applied low level laser therapy to the entire wound surface using her hand-held Resonator laser (an earlier version of the current and highly effective Q1000 model).

The horse's owner applied the low level laser every day for the first few days. Then she treated the mare every two days, and then once a week. After eight treatments, despite the severe infection and gangrene, the mare's leg began to heal. The next picture (Figure 6) shows the results after the wound healed. A huge, ugly scar remained and the horse limped, but the mare's life was saved. The concerned owner continued to treat her horse with low level laser therapy once a week for another three weeks

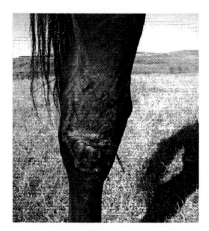

Fig. 6

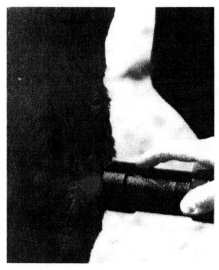

Fig. 7

The next picture (Figure 7) shows what happened after completion of the low level laser therapy. To everyone's amazement, the scar tissue on the horse's leg was soon covered by hair. Scar tissue does not usually regenerate hair. This hair growth demonstrated the thoroughness of the horse's recovery as a result of the regenerative powers of low level laser therapy and a special salve applied by the owner to keep the scar soft. You may wonder what brought about such marvelous healing. Quite simply, the cell membranes surrounding the wound were re-energized. This allowed osmosis to occur and the essential hair-growing nutrients already present in the horse's blood were allowed to work. The cell nutrients were restored to their normal state. More nourishment flowed into the tissues from the formation of new blood vessels, leading to changes in the cellular DNA, which eventually returned to normal. The owner's brood mare later went on to produce three healthy foals—thanks to low level laser therapy.

Quick Relief After a Hot Tub Accident

The next series of five photographs show what happened when a thirty-eight-year-old man burned himself in his hot tub. Unbeknownst to the man, the thermostat of his tub malfunctioned allowing the

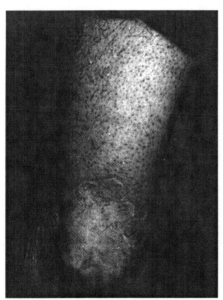

water temperature to reach 140 degrees Fahrenheit (140 F°). Upon stepping into the tub, he immediately removed his foot from the scalding water, but not before he sustained a painful second-degree burn on his lower leg. The first picture (Figure 8) shows how the water essentially dissolved some of the skin on the back of his foot and totally destroyed the top layer of skin on the shin and calf.

Fig. 8

Burns are difficult to treat medically because there isn't much that can be done. After assessing the extent of the damage, doctors can only wait for the body to heal itself. The damaged skin is very vulnerable to infection. Bandages not only interfere with the healing process by adhering to the wound but also provide an environment for breeding infection. Since human skin is loaded with nerve cells that generate pain impulses, burn pain is very intense. Anything that promotes healing also helps alleviate pain.

Fortunately, this burn victim received low level laser therapy right away. It turns out that he employs it as a routine treatment for all types of injuries at his home and at work. He borrowed a second laser and applied low level laser therapy over the entire surface area of the wound using both a hand-held Resonator laser and a Rotary Multiplex laser (both products a generation older than the current Q100 and Q1000 models.) He administered laser light from both instruments (because of the size of the injury) for two hours on the first day. A burn ointment containing Aloe Vera and ice was also applied.

The next picture (Figure 9) depicts how part of the wound

had turned bright red and a shiny glaze of severely damaged skin had formed. It is common knowledge among physicians that a serious burn usually requires four weeks to heal over. After the initial treatment, the length of time for the application of low level laser therapy was shortened each day.

Fig. 9

Fig. 10

The third picture (Figure 10) taken on the fourth day, shows the rapid regeneration of capillary tissue under the burned layer of skin. An ointment containing silver nitrate, known to protect against infection, was also

Fig. 11

applied. On the seventh day (Figure 11), the wound appears to have a purple tinge, indicating that more new skin growth containing the protein collagen had occurred. By this time the pain had become more bearable.

Pictures (Figures 12 and 13) taken on the ninth day, show that the burn is almost completely healed. This means that this severe, second-degree burn took about a third of the time normally required for healing. The damage remaining appears about as serious as mild sunburn. Moreover, hair is starting to grow back in the burn area. There is no scarring and, most importantly, no occurrence of infection.

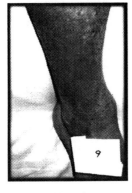

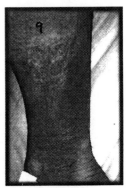

Biomedical researchers in Brazil recently conducted a laboratory experiment on biostimulation using low level laser light on cells called fibroblasts. They state in their published report, "Our results showed that a particular laser irradiation stimulates fibroblast proliferation, without impairing procollagen synthesis."[42] The burn victim's skin probably responded in much the same way as indicated in this experiment.

Fig. 12

Fig. 13

<u>Low Level Laser Therapy Heals a Surgical Wound</u>

After receiving surgery to correct an abdominal hernia, a sixty-year-old woman found that her navel wound, eight inches long and five inches wide, would not heal for an inordinately long period of

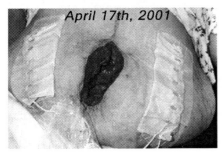

time (Figure 14). Figure 15 is a close-up of the same wound. Upon close examination, pus can be seen in the wound. After trying many medical approaches to no avail, Dr. Stephens, her surgeon, treated her with low level laser therapy. He applied

Fig. 14

the Rotary Multiplex laser for three minutes once daily for 16 days. Since the wound was so large (about the size of her foot) and the Rotary Multiplex laser emits a beam about the size of a small grapefruit, the laser had to be moved several times to cover the entire wound. As exhibited on the May 3rd picture (Figure 16), 16 days after the initial picture, the wound had healed to the size of a pecan. In another 16 days, it was nearly completely healed (Figure 17). Shortly after these pictures were taken, the wound had completely healed and the women's belly was normal for the first time since the injury.

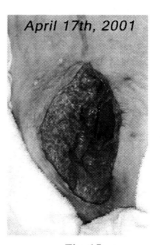

April 17th, 2001

Fig. 15

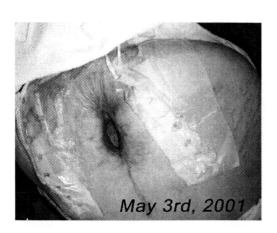

May 3rd, 2001

Fig. 16

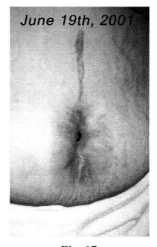

June 19th, 2001

Fig. 17

Low Level Laser Therapy Reduces Swelling and Speeds Healing After a Bicycle Accident

Eli, a 5 year old young cowboy, was more skilled at riding his horse than a bicycle. He jackknifed the front wheel and had a serious accident, striking the pavement face first, hard enough to knock his front teeth, including the bone that supported them, back into the roof

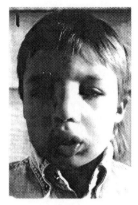

Fig. 18

of his mouth. Photograph 1(Figure 18) taken 5 hours after the accident shows considerable swelling and skin abrasions. Fortunately his mother owned the Resonator laser and applied it to Eli's face in three locations for two minutes each five hours after the accident.

On day two, just 24 hours after applying the laser, photograph 2 (Figure 19) shows that the swelling is completely gone, a scab is clearly formed and his cheeks were rosy. Because the laser is so easy to use, Eli applied it to his own face on the second, third, fourth and fifth days for just two minutes in the same three locations where his mother used it on the first day. To his mother's amazement, healing was so fast that the scab fell off the third day (Figure 20). I personally saw Eli on the sixth day (Figure 21) and I could not tell, after examining both the inside and out-side his mouth, that he had had an accident. He was completely healed!

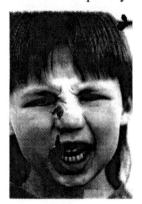

Fig. 19   **Fig. 20**   **Fig. 21**

As a practicing dentist for 35 years, I have seen many trauma accidents to teeth. This was the first time I have seen baby (primary) teeth receive such a severe blow and not turn black. Another big plus is that because his mother owned the laser and used it right after the accident, there was no damage to his permanent teeth. Since this accident, Eli has become an experienced bicycle rider as well as a horseback rider.

Low Level Laser Heals Tongue Lesion

These pictures and case history were submitted by Marjorie Orser, Registered Dental Hygienist and a regular Q Series laser user from Palm Springs, CA.

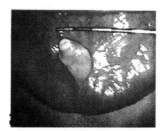

Fig. 22

A lesion under the tip of the tongue was found on routine examination on a 17-year-old male patient in April, 2004 (Figure 22). The patient was referred to an oral surgeon and the lesion was removed on April 04, 2004 and sent for laboratory evaluation. The pathology report diagnosed the lesion as a nonmalignant Mucocele.

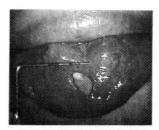

Fig. 23

On June 29, 2004, less than three months from the time of the surgical removal, the lesion was back again, only this time it was twice the size (Figure 23). Rather than repeat the surgery, the lesion was treated with low level laser therapy. The 660 Enhancer followed by mode 3 of the Q1000 was applied for three minutes on June 29, 2004. The patient received similar laser treatments on June 30th and on July 1st. In just two days the lesion had shrunk in size by over one half (Figure 24). Four more laser treatments were repeated on July 6th, 7th, 8th, and 9[th.]

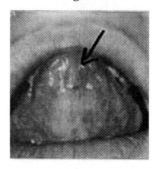

Fig. 24

The patient returned on July 19th and it was noted that in just 2 ½ weeks the lesion had totally healed (Figure 25).

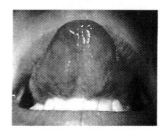

Fig. 25

While Mucocele lesions are not life threatening, they are annoying and worrisome to the patient and normally require painful surgery to remove them. This case history presents a better, more comfortable way to treat lesions of the mouth.

Because this method of treatment is so comfortable and safe, low level laser therapy should be considered as the first choice of treatment for any oral lesions or lesions on any other part of the body.

5.5   Research Indicates that Laser Light Can Reduce Bacterial and Other Infections

If you own or train horses or talk to any one who does, you probably understand that the healing of the mare's leg injury (Figure 7), to the extent that hair grew back on the wound, is akin to a miracle. Prior to laser treatment, we saw a weeping, pus-filled wound. However, by the second day of low level laser therapy, the owner reports that the animal's wound began drying up. The laser definitely appeared to produce an antibacterial effect. The same process occurred in the case of the above-described abdominal hernia.

But what do scientists say about this? Sanford Simon, Ph.D., Professor of Biochemistry and Pathology, Stony Brook Health Sciences Center, State University of New York in Stony Brook, studied the Rotary Multiplex laser (an effective predecessor to the present highly beneficial Q1000 laser) to determine the safety and efficacy of this type of low level laser. Dr. Simon states in his research report: "These lasers are non toxic to all cells. Certainly higher power cutting lasers can destroy bacterial cells, but the laser used on these wounds was very low power and does not kill cells." Rather, low level laser therapy does something else to pathogens, something that has proven to be remarkable.

All living creatures, including single cells such as amoeba, have a protective mechanism. It is believed that simple cells do not like laser light and move away from it. Low power lasers seem to create a reaction inside the infecting bacteria that cause them to move away. As a high school Biology teacher, I had my classes look at amoeba under a microscope to see how it slowly moved about in the medium. When my students stimulated this single-celled amoeba with a sharp pointed instrument, the amoeba slowly moved away from the stimuli and did not return to that part of the slide. I always wondered why.

Bacteria, multiplying within tissue wounds, appear to shy away from the laser light stimuli in the same manner. When the pathogens retreat to a region with a healthier blood supply, the body's defense system tends to destroy them. The important point here is that the person applying low level laser therapy controls bacteria in a natural way. No need to over load an already weakened body with antibiotic drugs, to which bacteria eventually develop resistances anyway.

Researchers conjecture that laser light probably works against viral infections in the same way. And there is speculation that the same process affects parasites as well. Just as amoeba move away from irritating stimuli, it is likely that fungi and other parasites respond in a similar way to lasers with the proper power density and frequency of laser light. Because it has already been confirmed that laser treatment is safe, it would make good sense to use it to counteract offensive attacks on humans, pets and farm animals by bacteria, viruses, fungi and parasites.

The body's immune system is constantly ridding the total physiology of invading bacteria. If allowed to multiply and spread throughout the body, such pathogens are potentially very harmful. An open wound infected by bacteria might never heal. By attending to the bacteria directly, the low level laser light provides a tremendous advantage to the immune system. **One could say that low level laser therapy tends to become a kind of "second immune system."**

While the manufacturer of the Q Series lasers does not claim any bacteria-killing effect for its products, nevertheless, exciting results are being seen throughout the world. For example, after irradiating marine bacteria with low level laser light, Japanese scientists recently reported "that low-power pulsed laser irradiations resulted in a significant bacterial mortality…"[43]

5.6   How Low Level Laser Therapy Promotes Soft Tissue Healing

Laboratory investigation has clarified how low level laser light promotes the healing of soft tissue in experimental animals. Researchers in Japan, experimenting on laboratory rats, found that low level laser irradiation caused dilation of arterioles.[44] Wider blood vessels allow more blood to circulate in the treated area. The resultant stimulation of microcirculation brings about faster, more complete healing.

Researchers in Russia may have found the reason why this increased microcirculation occurs. They observed that a particular wavelength of laser radiation could free nitric oxide (chemical symbol NO) from the hemoglobin molecule and thereby create free NO.[45] Hemoglobin, a molecule in red blood cells involved in transporting oxygen, actually regulates blood flow. It does so by changing shape and releasing a souped-up molecule of nitric oxide called s-nitrosothiol (SNO). SNO is carried by hemoglobin with oxygen through the blood stream. Thus, hemoglobin simultaneously releases SNO to dilate blood vessels and delivers oxygen to nourish tissues. When oxygen levels are high, hemoglobin scavenges excess oxygen and NO, constricting blood vessels and reducing blood flow. This is the model scientists have come up with to explain how hemoglobin transports oxygen.

Nitric oxide is one of the simplest molecules found in nature. It's a colorless and odorless gas composed of one atom of nitrogen and one atom of oxygen. NO is a crucial component in a wide variety of biochemical processes that take place throughout the human body. In 1998, physician and scientist, Ferid "Fred" Murad, M.D., Ph.D., co-winner of the Nobel Prize in Medicine/physiology for his work on nitric oxide,[46] identified that one of NO's functions is to signal the muscles that control relaxation and expansion of blood vessels. When muscles relax, the blood vessels widen and blood flow increases (vasodilatation). Laboratory experiments at Harvard Medical School using blood vessel tissue from rabbits also indicated that certain wavelengths of low-power laser radiation could trigger relaxation of blood vessels.[47] The Harvard scientists called this response "photovasorelaxation."

Another study showed that low level laser enhances the production of NO.[48] This starts a chain reaction which stimulates cyclic guanosine monophosphate (cGMP) synthesis, which in turn reduces intracellular Ca levels and relaxes the smooth muscle cells (SMC) of the Corpus Cavernosa allowing the arteries to fill with blood. While this was an *in vitro* study on penile smooth muscle cells, arteries are arteries regardless of their location. They all respond in the same way and it is good to have them filled with oxygenated blood.

The only difference between an artery and a vein is the smooth muscle layer between the inside and the outside layer of the vessels and the fact that arteries carry slightly more oxygenated blood. The inside layer of cells that line a blood vessel is called the "intima" and the outside layer of cells is the "exima." In arteries there is a smooth muscle layer between the intima and the exima. This raises the question, if the inside of our arteries and veins are basically the same, why doesn't plaque build up on the inside of veins? I asked a cardiologist this question and his answer was: "because it doesn't." As an ex-school teacher, I did not consider that a very good answer. It appears obvious that if the main difference between an artery and a vein is a smooth muscle layer, then something must occur in the smooth muscle layer to cause the plaque build up.

A better answer is that when the smooth muscle layer contracts and suffers miniature tears, fibrin is laid down as the beginning of a clot. This fibrin can eventually become arterial plaque when calcium, cholesterol and other blood components stick to it. These miniature tears may be related to an imbalance in nutrients such as vitamin C, CoQ10 and vitamin E, or to long chain polypeptides and macromolecules. In laymen's terms there is a problem with "incompletely digested food particles" or a combination of factors. When the long chain polypeptides, the high density lipoproteins (HDL) and the low density lipoproteins (LDL) get out of balance in the blood stream the smooth muscles contract and tear, leading to the formation of a clot. When smooth muscle constriction, an important defense mechanism, occurs throughout the body—the body goes into shock. When the Parasympathetic Nervous System (PNS) is not functioning properly and there is a deficiency of digestive enzymes, incompletely digested food particles or long chain polypeptides accumulate. This process is called inflammatory heart disease; the clots that result from the smooth muscle constriction are referred to as viable plaque. This is why plaque is only found in arteries where there is smooth muscle. One important thing to remember about low level laser therapy is that low level laser light releases tight smooth muscles quickly. This is another good reason to never leave home without your laser! It could save someone's life.

At a medical convention, I was explaining how low level lasers work to release smooth muscle and aid in treating heart attack. When I stated: "never leave home without your laser," a doctor stepped forward, took his laser out of his coat pocket and said: "you better believe it." On a recent trip, a flight attendant asked if there was a medical doctor on board. I volunteered and when I reached the front of the plane, a man was laying in the aisles. He was not breathing and had turned blue. Rather than perform CPR, I used my laser, first on the heart acupoint and then directly over the heart. Within ten minutes he was sitting in his seat breathing normally and his color was restored.

Medical researchers in Austria recently reported the same results with diabetic patients.[49] After applying low level laser therapy to the patients' feet, they noted: "Our data show a significant increase in skin circulation due to athermic [not heat generating] laser irradiation in patients with diabetic microangiopathy and point to the possibility of inducing systemic effects." This finding is significant because it indicates that low level laser therapy stimulates blood flow in those with impaired circulation. These finding are especially good news because diabetics are known to have peripheral circulation problems in their lower extremities that can become severe enough to require amputation of the toes, feet and lower legs.

Blood contains components of the immune system that fight infection. Blood also transports nutrients and raw materials the body needs to heal. For both these reasons, increased blood flow at the site of an injury, or in areas with impaired circulation, is especially beneficial for promoting healing. Laboratory research indicates that increasing microcirculation is one of the healing mechanisms triggered by low level laser therapy.

Dr. Walsh, in the department of dentistry at the University of Queensland, Australia, proposed that the following mechanisms (please see table below) are involved in the acceleration of wound healing by low level laser therapy.[50]

Possible Mechanism Involved in the Acceleration of Wound Healing by Low Level Laser Therapy

The following explanations may sound too technical for laypeople, but when you share this book with your doctor, he will appreciate the research behind low level laser therapy.

The laser light activates fibroblasts, the most important cells involved in wound healing, to proliferate and mature. It also causes the fibroblasts to move more actively, transform into myofibroblasts and secret more Basic Fibroblast Growth Factor (BFGF). Laser light also has positive impact on other cells that are involved in wound healing, as shown below.

Fibroblasts

- Proliferation
- Maturation
- Locomotion
- Transformation into myofibroblasts
- Reduced secretion of PHGE2 and IL-1
- Enhanced secretion of BFGF

Macrophages:

- Phagocytosis (uptake of broken down cell debris)
- Secretion of FGFs
- Fibrin resorption

Lymphocytes:

- Activation
- Enhanced proliferation

Epithelial cells:

- Motility

Endothelium:

- Increased granulation tissue
- Relaxation of vascular smooth muscle

Neural tissue:

- Reduced synthesis of inflammatory mediators
- Maturation and regeneration
- Axonal growth

This is an outline of how wound healing occurs. In summary, wound healing requires the participation of fibroblasts, macrophages, lymphocytes, epithelial cells and the endothelium. They all do their jobs and work in harmony to heal the wound. Experimental studies have shown that low level laser therapy stimulates these components and accelerates the process of healing.

5.7   Low Level Laser Therapy Promotes Hard Tissue Healing as Well

Anecdotal and scientific evidence demonstrating that low level laser therapy stimulates bone regeneration has been reported over the past 20 years. The effects, both local and systemic, include contra-lateral effects. Reports of accelerated healing of rabbit radii fractures and mouse femurs were made as early as 1986 and 1987. The irradiated bones healed faster than the bones of non-irradiated animals. Interestingly, contra-lateral (opposite side but same position) non-irradiated fractures healed as quickly as the irradiated bone, indicating the occurrence of a systemic effect. This study is further evidence that low level laser therapy has a cumulative and cascading effect. It is beneficial regardless of where it is introduced into the body. To date, much of the supporting evidence showing the benefit of low level laser therapy comes from experimental animal studies, while the number of human studies still lags behind. Most research first relies on animal experiments. Nevertheless, much initial evidence already reveals some benefits to humans. I am very hopeful that further research will soon demonstrate even more benefits to humans.

In various cell cultures and animal models, it has been clearly shown that low level laser therapy stimulates osteoblast cell proliferation and differentiation. Low level laser therapy-induced proliferation enhances bone nodule formation. Low level laser therapy-induced differentiation results in an increase in the number of more differentiated osteoblastic cells and an increase in bone formation. Low level laser therapy also induces the collagen fiber lamellar organization in the bone matrix. This effect is indicative of bone maturation.

Researchers from Italy, Brazil, Japan and other countries have contributed to the field of bone regeneration research.[51, 52, 53, 54, 55, 56, 57, 58, 59, 60] (Little research is being conducted in the U.S.) In addition, supporting evidence on the affect of low level laser therapy in bone regeneration comes from the area of dental research.[61, 62, 63, 64, 65, 66] Bone regeneration and implant integration in particular, are frequently the subjects of research. These studies show that low level laser therapy accelerates the growth of bone that holds the implants in place. Moreover, in a study of wound healing following tooth extraction in a rat model, daily low level laser therapy increased fibroblast proliferation and accelerated formation of bone matrix.

A similar canine model study failed. However, in that study low level laser therapy was given every other day and the irradiation levels were low. Thus, the negative finding may not mean that low level laser therapy is ineffective, but simply that the study design was poor. It is most likely that the delivery systems refining specific wavelengths, duration and frequency of treatments, as well as the power density have to be better characterized to increase the efficacy of low level laser therapy in hard tissue regeneration.

Of course, there is less available evidence supporting the efficacy of low level laser therapy for bone regeneration in human studies. However, W. R. Bennett's studies demonstrated that micro currents could enhance bone growth in frogs and humans.[67] Although the experimental evidence accrued from *in vitro* and *in vivo* studies suggests that low level laser therapy will be able to positively effect bone regeneration in humans, the hard proof concerning the specifics is still missing.

5.8 Low Level Laser Therapy Reduces or Eliminates Inflammation

Inflammation is the body's common response to disease and injury. It manifests as tissue redness, heat, swelling, and pain. In some acute traumas, inflammation initially acts as a protector. However, in the presence of chronic disease, inflammation and its damaging by-products are the body's enemies. For example, swelling, that accompanies chronic inflammation, chokes off blood flow and prevents an adequate nutritional supply to reach the damaged area. Pain, often associated with chronic inflammatory disease, can be terribly debilitating.

Several inflammatory enzymes released by damaged cells cause the pain, swelling, heat and redness of inflammation. When I was in China studying low level lasers, I heard a Chinese doctor present his research. He had conducted experiments on rats and proved that low level laser therapy tends to reduce or eliminate seven out of the nine inflammatory enzymes by as much as 75 percent. The same Chinese investigators have shown another benefit of low level laser therapy for inflammatory disorders—it actually enhances the production of healing enzymes. The photographs in (Figures 18-21) illustrate how low level laser therapy helped reduce inflammation in the little boy's face after the bicycle accident. The boy's illustrated case history is a fine example of how the Resonator Laser (now the Q100) reduces pain and inflammation and definitely speeds up overall healing.

At this writing, the manufacturer of the Q Series laser group (Q100 and Q1000) has research underway to show how low level laser therapy reduces pain and inflammation in osteoarthritis of the hand, knee, and hip. Clinically, the Q Series laser group has already demonstrated effectiveness for eliminating arthritic pain and inflammation.

5.9   A Word of Caution Concerning those Negative Studies of Low Level Laser Therapy

There are some negative studies of low level laser in the literature. My comment on these studies is simple; they did not get the wavelength, dosage and power density right. These factors will affect the outcome of the studies. This argument is supported by the authors of *The Low Level Laser Therapy Internet Guide,*[68] who analyzed a number of frequently cited negative studies on the effects of low level laser. Their analysis revealed that the most common reason for the negative result was the use of the wrong laser or the wrong dosage. Other reasons include faulty inclusion criteria, inaccurate control group definition, ineffective methods of therapy, inadequate attention to systemic effects and tissue condition, and low or varying power density. Another weakness often seen in these studies is their failure to provide sufficient data on laser parameters.[69]

The proper control of the wavelength and the joules of energy delivered are essential. DC batteries that become run down and depleted often cause the joule level, the output of energy, to vary. This could affect the clinical outcome. Because of this problem, the Q Series lasers use a sophisticated computer system to regulate the rechargeable batteries so that they deliver the proper power density, exact wavelength, exact frequency and exact joules of energy each time the laser is turned on. Any laser that does not control these aspects of laser physics to the "nth degree" will not be as effective. Since negative studies are often quoted as "proof" of the ineffectiveness of low level laser therapy, it is important that these studies be subjected to a proper critical analysis.

The Cochrane Collaboration, an international nonprofit group that evaluates research about clinical practices, has published several reports on low level laser therapy. The group found that low level laser therapy was effective in reducing pain and morning stiffness from rheumatoid arthritis, as well as frozen shoulders. But it did not find benefit in treating osteoarthritic pain and rotator cuff tendonitis.[70, 71] However, two studies have shown very positive results for the treatment of osteoarthritic pain with low level laser therapy. According to Lars Hode, a Swedish physicist and the president of the Swedish Laser-Medical Society for the year 2002, the safety and efficacy of low level laser therapy is better documented than that for ultrasound therapy, which is well accepted in the medical world. And, he further explains, low level laser therapy is much improved today and should not be held back by those negative studies that are more than 20 years old. [69]

After reading about low level laser therapy, you might come to the conclusion that it sounds "too good to be true." However, all the positive double blind studies and accumulating patient testimonials clearly demonstrate the growing potential of low level laser therapy for alleviating many health problems. It may not cure or completely reverse all diseases; but some relief from chronic and acute problems, with a simple hand-held, non-invasive device certainly is a great addition to anyone's home care arsenal. Before you purchase a low level laser, you can get treatments for your indicated condition to test out its efficacy for yourself. The small investment in these treatments is rather affordable.

Finally, Lois Lindstrom, the author of *Memoirs of a Swedish War Nurse*, wrote a special report for the February, 2004 *Washington Post* about the use of cold laser therapy by sports teams.[72] Interestingly, she reported that the New England Patriots might have won the Super Bowl XXXVIII with the help of a cold laser. According to Lindstrom, more than 10 Patriot players were treated with cold lasers during the week before the Super Bowl for tendon and muscle injuries. Ellen Spicuzza, the physical therapist who treats the Patriot players, added low level laser therapy for this big game for the first time. Usually, she treats player's injuries only with medical massage. But now,

Spicuzza believes that low level laser therapy helps the injuries of the Patriot players. She also uses laser for her own ten-year-old ski accident injury and gets quick relief from muscle spasms. There is no doubt that athletes and sports medicine doctors are catching on to cold laser therapy because it heals injuries faster. Much is at stake for players with multi-million dollar contracts. You can be certain that good sports doctors will find treatments that can quickly heal their patients. The U.S. Olympic training centers in Colorado Springs, Chula Vista, California, and Lake Placid, New York, all offer low level laser therapy to their athletes.

Chapter 6   Applications of Low Level
Laser in Acupuncture

6.1   East Meets West

Laser therapy can be integrated with other healing disciplines as well. This chapter presents an introduction to Traditional Chinese Medicine (TCM) and the evolution of TCM in the United States, leading to its acceptance as a legitimate healing tradition. Then we will focus on a medical doctor who has integrated low level laser therapy and acupuncture in his practice. Low level laser therapy, a modern invention, harmonizes well with the ancient healing modality, acupuncture; they both aim to enhance the body's built-in ability to repair itself

6.2   The Theory of Traditional Chinese Medicine:
A Different Medical Paradigm

The logical structures underlying the practice of Traditional Chinese Medicine and Western medicine are radically different. The Chinese see the complete physiological and psychological individual—body, mind and spirit—as being fully integrated aspects of human life. During the diagnostic process, "all relevant information...is gathered and woven together until it forms what Chinese medicine calls a 'pattern of disharmony'...or a situation of imbalance in a patient's body."[73] There is no requirement to identify a specific disease or a precise cause. Illness is not perceived as a collection of symptoms we would like to make disappear. Rather, "the Chinese are interested in discerning the relationship among bodily events occurring at the same time...the therapy attempts to bring the configuration into balance, to restore harmony to the individual."[73]

The basic theoretical concepts of Traditional Chinese Medicine, which differ from those of western medicine, are: The Five Elements, *Qi* (pronounced Chee and often spelled Ch'i or Chi) and *Yin-Yang*.

The Five Elements

The traditional Chinese, like many ancient societies, saw human beings as completely enmeshed in nature. The interaction of the Five Elements, of which nature and humans are a part, brings forth universal harmony. The Five Elements—wood, fire, earth, metal and water—are linked with all other things, such as the cycle of the seasons, the time of day, the colors, the bodily organs, the directions, the flavors, the emotions, and the list goes on. Chinese doctors take these factors into account when working with patients.[74]

The Energetic Life Force: "*Qi*"

Thousands of years before they were exposed to the knowledge of modern medicine, the Chinese developed a traditional form of medicine (TCM) based on the theory that the energy (*qi*) circulating through the human body is responsible for coordinating the bodily functions. According to Chinese physicians, the same *qi* that fills the universe also moves through our bodies. This qi energy is the life force responsible for generating metabolic processes. Every health problem can be traced to a problem with the patient's *qi*.

Qi is also the central principle in *Qigong* (*qi* cultivation practice) and *Taiji* (*Tai Chi,* the slow moving meditative practice that many of us have seen). Using bodily movements and mental concentration, *Qigong* and *Taiji* were developed in China, along with acupuncture, as a means of stimulating and focusing *qi*. In fact, all of the Asian martial arts, whether from Korea, Vietnam, China or Japan, operate on the basis of *qi*. In Japanese, the word *qi* is pronounced *ki*. For many practitioners, *qi* energy is a philosophy that guides them through life. In fact, a good number of Traditional Chinese Doctors are also *qi* masters, as are many painters and calligraphers.

The theory encompassing the presence of *qi* is very appealing for two reasons. First, although biochemists can describe biological systems according to their scientific understanding of natural forces and life processes, they cannot identify precisely what "life force" drives all the chemical reactions occurring in the body. That is, they can describe what happens, but they cannot explain why it is happening. For example, what happens when a cell dies? The life goes out of it and the matter stays behind, but what does this really mean? The concept of *qi* is the beginning of some sort of answer to this question. But, the concept is very appealing for another good reason too. The Chinese medical tradition, built around *qi*, has achieved excellent results.

The formulation of a theory, and the discovery of evidence that corroborates the theory, are normal features of scientific research. It is foreseeable that a method will eventually be found to detect and measure *qi*, thereby proving its existence. For now, it is reasonable to consider it a working hypothesis, especially in light of the success the Chinese, and others, have had in using it for treating patients.

Good Health Requires the Balance of *Yin* & *Yang*

Figure 26

The natural balance so important in Traditional Chinese Medicine arises from the fundamental theory of *yin and yang*, which states that life takes place in an alternating rhythm of opposites, such as dark and light, male and female, left and right, in and out, up and down, and yes and no. Flowers open and close, the moon waxes and wanes, the tides come in and go out. We wake and sleep, breathe in and breathe out. *Yang*, the male energy, is characterized by the daytime and the season of spring, both periods of light and increased activity.

Yin, the female energy, is the darkness of night, the turning inward and the winter resting phase. Y*in-yang* is a constantly changing, continual flow through which everything is expressed on the one hand and recharged on the other. They are a unified couple. Their proper relationship is health; a disturbance in this relationship is disease.[75]

Although we can define *yin* and *yang* individually, and distinguish one from the other, they are inseparable. Wherever there is one, there is also the other. The symbol of y*in-yang*, a circle separated into a white half and a black half has become known universally (Figure 26). It demonstrates that nothing in the universe is pure *yin* or pure *yang*; within this symbolic circle, black and white embrace and intertwine in perfect symmetry, each side containing a small circle of its opposite.

Obviously, the concept of *yin* and y*ang* is hardly how scientists explain their theories in Western civilization; but do not be misled by the poetic sounding descriptions. During some five thousand years Chinese physicians, using hands-on experimentation plus much trial and error, have developed quite specific medical techniques based on the general principles of *yin* and *yang*. The techniques have proven themselves repeatedly over the centuries. Like Western scientists, the Chinese healers have published their results in long treatises for their fellow practitioners. Thanks to their diligence and professionalism, these texts and techniques have endured over time. We, in the Western industrialized countries are the modern-day beneficiaries and should appreciate the ancient Chinese healing masters.

6.3 Acupuncture—An Important Component of Traditional Chinese Medicine

Acupuncture is the most well known feature of Traditional Chinese Medicine in the West, but to the Chinese, herbs and *tui-na* (massage) are also important components of the system. All three branches of Traditional Chinese Medicine aim for the same goal—balance in the mind-body-spirit of the individual.

Acupuncture, the Science of Meridians and Acupoints

When a proper balance of Yin-Yang forces exists, the body has achieved a healthy circulation of *qi,* which flows through 12 main meridians or pathways with more than 362 acupuncture points or acupoints. The activation of these points by needling serves as a way to balance the *qi* so that the body can heal itself. When the *qi* flow is insufficient, unbalanced, interrupted or blocked then the application of acupuncture is in order. Acupuncture can be described as the insertion of very thin metal needles resembling segments of fine wire into the skin at the points. Some points move *qi* towards the interior of the body, while others bring *qi* to the surface.

Traditional Diagnosis

The choice of acupuncture points varies from patient to patient and treatment to treatment and relies on very careful diagnoses. Chinese doctors diagnose patients by looking, listening, smelling, asking and touching. "Looking begins as soon as the patient enters the office; the doctor notices their vitality, the color of their skin and clothing, their posture and their eyes. He also, and most importantly, examines the tongue, the only place inside of the body that is seen. Listening gives the patient a chance to detail their story and symptoms. As the story is told, the practitioner listens for the emotional quality of the voice, the flow of the breath and where the tension shows. Smelling can give clues about the patient's health in relation to the Five Element cycles. Very curiously, the written character for listening and smelling is the same.

The next part of gathering information for diagnosis is asking. Ten questions have been designed to learn the most specific patterns presented by the patient. These include questions about perspiration, hot and cold, fluids, diet, digestion, urination, bowels, pain, emotions, sleep, and for women, gynecology. This information gives a picture of how the body functions in relation to *yin-yang,* the quality of *qi* in all areas, and the interaction of the organs and elements.

The last part of the diagnosis, touching, includes the important act of pulse taking. There are three pulses on each wrist with at least two levels at each pulse. That makes twelve pulses, each reflecting the state of health of one organ. There are twenty-eight qualities of the pulse that further describe the history of the organ and the flow of *qi* and blood. This information is added to the patient's picture.

The diagnosis is not biomedical. Rather, it offers a description of the flow of energy, life force, through a dynamic whole organism. And by reaching the diagnosis, the treatment is defined and the patient educated as to the patterns and causes of the imbalances in their body."[76]

After thousands of years of observing the effects of acupuncture treatment, the Chinese have developed visual diagrams showing the routes of the meridians through the body and the location of the points. They have also determined which points are involved with particular disease symptoms and the combination of points necessary to treat certain conditions. The needles being inserted are sometimes combined with electrical stimulus referred to as "electro acupuncture" or with heat produced by the burning of special herbs in a procedure called "moxibustion." This entails the attachment of a burning ball of herbs to the top of the needle after it is inserted into the skin.

6.4   The Development of Acupuncture in China

Though historians are unsure of its exact age, the Chinese healing art of acupuncture can be dated back at least two thousand years and some historians maintain that it has been practiced in China for four thousand years. Several medical texts were produced in China during the Early Imperial era (220-960 AD). These early documents trace the development of the major conceptual features and theories of Chinese medicine, which include anatomy, physiology, and pathology. One particular text, the *Huang-di Nei-jing* (*Inner Classic of The Yellow Emperor*), probably compiled around 100 B.C., is the cornerstone of acupuncture. This systematic medical textbook contains two sections: the first section, the *Su Wen* (*Fundamental Questions*) discusses medical theory; the second section, the *Ling Shu* (*Spiritual Axis*), is the fundamental acupuncture manual. These two texts together explain the Yin-Yang theory. The underlying theme is that individual symptoms are somatic (bodily) processes rather than supernatural events. By the time the *Huang Di Nei Jing* was written, many of the acupuncture meridians (channels) and points had been identified. Also, the theory of the circulation of *qi* through the meridians became the dominant concept of Chinese medicine. *Qi*, was described as a product of both the body and the environment. Illness resulted when "healthy" bodily *qi* was disrupted or when "evil" external *qi* entered the body. With its basic practical and philosophical tenets in place, acupuncture continued to develop in China over the next 2,000 years.

Chinese acupuncture reached a higher level when, in 282 A.D., a medical scholar named Huang-fu Mi completed the *Zhen-jiu Jia-yi Jing (Systematic Classic of Acupuncture,* sometimes translated as the *ABC of Acupuncture*). It is a systematic presentation of material from the *Ling Shu* and other early, but now lost, texts.[77] In it he identified the currently known acupuncture points and offered detailed descriptions of the meridians, naming and listing the location of the points on each, and explaining how deeply each should be needled. The text also records the length of time to leave the needles in, the number of moxa cones to be applied to each point, and what each point is known to treat. This is the first text to advocate acupuncture as not only a cure, but also a means of disease prevention. It emphasized that with the proper knowledge, the patient and physician can treat disease before it arises. This is a cornerstone of Chinese medicine—unlike western medical doctors, Chinese physicians traditionally got paid for preventing disease, not for treating it. The *Zhen-jiu Jia-yi Jing* is a milestone in the progress towards perfecting the healing art of acupuncture. Many of its ideas continue to be used to this day.

The spread of Buddhism to the neighboring lands of Japan, Korea and Vietnam carried the knowledge of Chinese medicine, including acupuncture, in its wake and so contributed to the practice of Traditional Chinese Medicine. The fundamental texts of acupuncture were particularly studied in Japan with great care. Universities in Japan support the scientific study of *qi* energy to this day.

It is noteworthy that Traditional Chinese Medicine and acupuncture remained the standard system of healing in China, and throughout Asia, for two thousand years. Only with the infiltration of western scientific medical ideas in the twentieth century did Traditional Chinese Medicine begin to suffer a severe decline. By the 1930's, traditional Chinese doctors were a disappearing breed, at least in the urban areas. It wasn't until the establishment of The People's Republic of China in 1949 and Mao Zidong's Cultural Revolution in the 1960's, that some important aspects of traditional Chinese medicine were revived. Eventually, interest in Traditional Chinese Medicine found its way to the United States in the 1960's when President Nixon visited China to open the US – China trade relationship. Since Nixon's visit, Traditional Chinese Medicine has greatly contributed to the growing movement of Complementary and Alternative Medicine.

6.5   Acupuncture In the United States

Acupuncture, now recognized as one of the important Complementary Alternative Medicine (CAM) therapies, is taught and licensed for practice in the United States. There are many Traditional Chinese Medicine and acupuncture schools located throughout the country; a good number of them are accredited. Acupuncture practitioners are not required to be physicians, but many medical doctors do learn some acupuncture. There are no less than four national medical organizations devoted to acupuncture therapy. There's no doubt that low level laser therapy applied to acupressure points will also beget its own professional health organization. American acupuncturists also subscribe to at least two medical journals, *Acupuncture Today* and the *Qi Journal*.

Acupuncture has demonstrated its usefulness in treating a wide variety of ailments, including life-threatening illnesses, as shown in the government studies listed below. It is also used to treat psychological conditions and addictions, including addiction to drugs and smoking. In the United States, acupuncture is considered to be a highly important CAM modality or holistic therapy. This means that it does not interfere with other forms of medical treatment and can be used in conjunction with pharmaceutical therapies and surgery. The mainland and Taiwanese Chinese, who developed acupuncture, have also found this to be true.

Acupuncture as a Form of Anesthesia

The Chinese use of acupuncture as a form of anesthesia became well known to Americans at the time of President Nixon's famous trip to China in 1971. One of the reporters, James Reston from the *New York Times,* required an emergency appendectomy and was anesthetized with the use of acupuncture. Later, Reston wrote about a new medical discovery called "acupuncture anesthesia."[78] Chinese patients often undergo major in-hospital surgical operations while fully conscious by having their pain blocked with acupuncture. This is especially important for patients who are allergic or sensitive to anesthetics, especially older people. The pain associated with dentistry in the United States is also being alleviated for those Americans consulting laser-enlightened dentists who employ low level laser therapy to administer acupuncture.

Medical Studies of *Qi* Conducted By the United States Government

Interest in Traditional Chinese Medicine and the theory of *qi* energy has been increasing in the United States for the past ten years among both patients and professionals. During the 1990's, the National Institutes of Health (NIH), an agency of the United States government, conducted many studies of *qi.* Some of them involved acupuncture and some investigated the medical uses of Chinese herbs, *Qigong* and T*aiji.*[79]

Some of the medical studies that investigated the influence of Traditional Chinese Medicine on acute and/or chronic physical, mental, and emotional illnesses sponsored by the United States government are the following:

- Qigong and Late-Stage Reflex Sympathetic Dystrophy (1993);
- Acupuncture Point Treatment for ADHA (1993);
- Taiji for Balance Disorders (1993);

- Acupuncture Treatment for Unipolar Depression (1994);
- Acupuncture Effect on Osteoarthritis (1994);
- Acupuncture vs. Placebo for PMS (1994);
- Acupuncture and Postoperative Oral Surgery Pain (1994);
- Acupuncture and Herbal Treatment of Chronic HIV Sinusitis (1994);
- Chinese Herbs—an Alternative Treatment for Hot Flashes (1994);
- Chinese Herbal Therapy for Common Warts (1994);
- . Laser Acupuncture Treatment for ADHD (1997);
- Acupuncture in the Treatment of Depression (1997);
- Neurobiology of Acupuncture (1997).

Many of these studies listed were conducted at hospitals associated with medical schools throughout the United States.

As a result of their ongoing research, in 1997 the NIH issued a consensus statement that found "promising" results for the efficacy of acupuncture in treating adult post-operative and chemotherapy nausea and vomiting, and in treating post-operative dental pain. It also stated that acupuncture may be useful in treating addiction, stroke rehabilitation, headache, menstrual cramps, tennis elbow, fibromyalgia, myofascial pain, osteoarthritis, low back pain, carpal tunnel syndrome, and asthma.

In fiscal year 2001, the NIH budgeted $1.5 million for the study of acupuncture. The agency is requesting that scientists conduct interdisciplinary research into the following:

- elucidating the basic biology and biochemistry of acupuncture;
- characterizing the actions of acupuncture at both cellular and systemic levels;
- defining actions of acupuncture on the endocrine, immune, and nervous systems, including any interactions between these three systems;
- identifying the biological and psychological sources of individual differences in response to acupuncture therapy, including genetic variations;
- defining and characterizing the biological basis for

individual acupuncture points;
investigating the scientific basis for key traditional
acupuncture concepts such as *qi* and the meridian system;
• developing objective methods to overcome the technical
challenges faced when studying the efficacy of
acupuncture.

In other words, the NIH would like to know how acupuncture works, whether it will work on Americans as well as it does for Chinese people, and if the *qi* energy can be detected and measured by some sort of electronic testing equipment. Scientists at Japanese universities are conducting the same sort of research.

National Center for Complementary and Alternative Medicine Conducts Studies on Chinese Medicine

The U.S. Congress authorized the formation of the Office of Alternative Medicine, one of twenty-seven institutes and centers that make up the National Institute of Health (NIH) in the early 1990s. The name of the agency was changed to the National Center for Complementary and Alternative Medicine (NCCAM) in 1998. This governmental agency was organized for the purpose of investigating newly developed alternative medical advancements and rediscovered traditional medical practices. NCCAM is currently sponsoring medical research on the effectiveness of acupuncture for treating seventeen major illnesses.

The NCCAM research includes the following:

• a study to see whether acupressure (acupuncture using
pressure applied by the hands instead of needles) can help
nausea and vomiting in persons with HIV/AIDS;
• a study to test the effectiveness of electro-acupuncture
as a treatment for hypertension (high blood pressure)
by evaluating treatment results according to Western
medicine benchmarks; a study to test treatment protocols
using acupuncture and moxibustion on patients with HIV
experiencing chronic diarrhea (which affects over 60% of
all HIV victims.);
• a study focusing on the use of acupuncture as a mode

of therapy for fibromyalgia (the second most common rheumatic disorder, affecting approximately 8-10 million persons in the U.S., characterized by widespread musculoskeletal pain and soft tissue tenderness);

- a large randomized placebo-controlled trial for testing the ability of acupuncture to treat major depression— evaluating treatment effects from the perspective of Western psychiatry.

NCCAM is also researching:

- the effectiveness of acupuncture in treating acute low back pain,
- cardiovascular disease,
- carpal tunnel syndrome (from repetitive stress disorder),
- spastic cerebral palsy in children,
- chemotherapy-induced nausea and vomiting in child cancer patients,
- colon cancer,
- reducing postoperative dental surgery discomfort (wisdom tooth extraction),
- endometriosis-related pelvic pain (caused by fibrous tumors on female reproductive organs),
- osteoarthritis of the knee, and
- Tempromandibular joint (TMJ) disorders (pain in the jaw).

Hopefully, NCCAM will soon begin studying the results from combining acupuncture and low level laser.

6.6 Replacing Acupuncture Needles With Laser Light

Normally, a patient cannot administer acupuncture on himself. Even if he could locate the correct meridian and acupuncture point, and assuming the point is located on a part of the body that is accessible, it would require an extreme amount of skill do it properly. This problem has been solved by the Q-Series lasers.

After years of research, I have devised a small hand-held diode laser that can stimulate the acupuncture points with a beam of light tuned to a specific frequency that benefits the human body.

This type of laser application is referred to as the administration of low level laser therapy. This laser device can concentrate laser light at the acupuncture points or at a broad expanse of multiple acupuncture sites on the body, depending upon which laser model is being employed.

There are many books on the market that illustrate the traditional Chinese acupuncture points. The *Low Level Laser User's Manual*,(to order call 605-342-5669) written to be used with the Q Series lasers, also includes a section that shows the points that can be activated with low level laser light as a substitute for needles.[80] The procedure for stimulating acupuncture points with laser treatment is very easy. The Q Series 660 Enhancer laser therapy instrument can be used by anyone. One points it in the area of the acupuncture point, and switches on the laser for thirty to forty-five seconds. The comforting therapeutic and preventive medical effects that result often amaze practitioners who administer low level laser therapy on their clients, as well as those who use it on themselves.

The Chinese and other Asian healers readily accept the use of low level laser therapy on acupuncture points. Their minds are open to new technologies which can be combined with ancient healing methods. The joining of acupuncture points with laser light is the ultimate example of the benefits to be derived from the integration of modalities and technologies from different traditions. In Japan, cold or soft lasers have been approved for treatment since 1987 and are widely used today.

6.7 Romeo Quini, M.D.: an Example of the Integration of Laser and Acupuncture

Romeo Quini, M.D., a family practice physician, received his medical training in the Philippines and undertook two residencies in general and thoracic surgery at Mercy Hospital and Stritch Loyola Medical School in Chicago. He learned classical acupuncture and herbology from a traditional Chinese doctor, and has utilized acupuncture in his San Diego, California medical practice since the early 1970s. A decade later, Dr. Quini began to use low level laser light to replace the "needles" for delivering his acupuncture treatment. His results for patients have been superb.

In a March, 2003 testimonial, Dr. Quini states: "I have been using Dr. Larry Lytle's low intensity laser in my medical practice for nearly two years. I utilize the principles of acupuncture points. They work in a fantastic manner for most medical ailments! As a physician I have used low level laser therapy for many medical conditions with very good results. The conditions include: trauma (hematomas resolve quickly, and sprains and strains find almost instant relief); arthritic pains (my patients achieve fast relief when some cases cannot be helped by conventional medications); and post operative pains. As a surgeon, I have used low intensity laser for control of many types of pain. Many of my patients do not need pain medications because of my application of post-operative low level laser therapy. They also recover faster."

In a general statement about his application of laser light to acupoints, Dr. Quini states: "Low level laser therapy, and acupuncture utilizing low level lasers have been an important part of my medical practice since 1980. The combination of acupuncture applied with laser light as the needling instrument has proven quite effective in pain control."

"I have recently applied low level laser therapy to acupuncture points on five postoperative patients, one who suffered from removal of their thyroid glands (thyroidectomy), one individual who had undergone surgical removal of the gall bladder (cholecystectomy), two persons who had had their appendices removed (appendectomy) and numerous benign soft tissue lesions," says Dr. Quini. "Low level laser acupoint therapy can provide very effective pain control in postoperative and medical conditions, in a non-invasive, cost effective manner. Utilization of this technique is easy, simple and can be readily learned by anyone."

"One of the appendectomy patients was a six-and-a-half-year-old child with a ruptured appendix," explains Dr. Quini. "Two days after surgery, the child was unable to move because of severe abdominal pain, and was dehydrated and unable to take oral fluids. I applied the low level laser at specific acupoints in the morning. By the afternoon of the same day, the child was out of bed walking with very little discomfort, and he was also able to take fluids."

"A farmer in his mid-thirties had been feeling exceedingly uncomfortable from the pain of herpes zoster lesions (shingles) on the right chest. He had problems breathing and could not move because of really severe shingles pain–despite taking pain medications.... I saw him as a patient for the first time when he consulted me for the elimination of his early onset vesicular lesions along the nerve trunks on the right chest. Low level laser therapy was given to the appropriate acupoints that same day and another laser treatment was delivered to the man early the next day. Without his taking any pain medications, after the second treatment the farmer felt only negligible pain and he returned to his farm to work. At that same time, the herpetic lesions [blisters] started drying up, which is quite astounding because it was so early in this case."

"I have also applied low level laser therapy to acupuncture points quite effectively in multiple cases of painful arthritis, headaches of various nature (including migraine), ankle and wrist sprains, strains, whiplash injuries, carpal tunnel syndrome and others," concludes Romeo Quini, M.D. "I rely on Dr. Larry Lytle's patented Q Series of laser instruments and turn to them to assure beneficial effects for my patients' health difficulties."

Each year over one million patients receive acupuncture in the United States. This amounts to approximately ten million treatment visits. These people are seeking relief or cures for a variety of medical problems, including addictions, allergies, chronic pains, headaches, and mental depression, to name just a few. Now that it has received the recognition it deserves, ordinary people seek out acupuncture on their own initiative or based on advice from other patients and the media. A typical acupuncture patient suffers from an illness that has already been treated by their regular family physician, but to no avail. They eventually may find relief with acupuncture, which sets in even more quickly when low level laser therapy is administered to the acupuncture points.

And the beauty of low level laser acupoint therapy is that it is simple to use by laypeople in the privacy of their own home.

Preface to Chapters 7, 8, 9 and 10

A reader might ask the question, "If *Healing Light* is a book about low level lasers, why are there several chapters dedicated to the Autonomic Nervous System, Dental Distress Syndrome and Proprioceptive Feedback to the Brain?" The answer is simple. The body is a unified energy system with interdependent parts.

Early records of Eastern medicine have stated that the teeth regulate the meridian systems. A book about healing light would be incomplete if it led the reader to believe that low level lasers are miracle devices that can "fix everything" without addressing underlying imbalances and root causes. Western medicine is a "cause and effect-search for a cure" type of medicine—viruses cause colds, bacteria cause infections, and the list goes on. Eastern medicine is more concerned with energy imbalances and is based on the belief that if the energy is in balance, the body can combat harmful bacteria and viruses.

The rest of this book will explain how the overall balance of energy in the body is linked to an often over-looked dental problem—the relationship of the lower jaw to the skull. Dentistry today operates on a fill-em, bill-em, clean-em, and when that fails, pull-em and plate-em type of dentistry. The important role that dentistry plays in regulating the Autonomic Nervous System and balancing the Sympathetic and Parasympathetic Nervous Systems is often ignored. Yet, this is the "missing link" in our pursuit of wellness.

(For more detailed information on this important subject, attend one of Dr Lytle's Healing Light Seminars – visit www. laserinformation.com for location or call 605-342-5669 to register)

Chapter 7   The Nervous System, Proprioception and Dental Distress Syndrome

7.1   Dental Distress Syndrome (DDS): a Significant Hidden Risk Factor in Disease

Stress is a significant risk factor in all disease, and that includes diseases of the heart. Hans Selye, the seminal thinker who first identified the importance of stress, has called Dental Distress Syndrome one of the worst causes of stress known to mankind.[81, 82] This "hidden stress" is estimated to affect over 75% of our population. A 1986 Harvard publication studied the morbidity (cause of death) of 33,000 doctors. As in the general population, the number one cause was heart disease and the number two cause was cancer. Surprisingly, the highest risk factors were not cholesterol, high blood pressure, lack of exercise, and obesity, but rather the loss of teeth. In fact, the loss of 10 or more teeth resulted in a 67% increase in heart attack and stroke. Moreover, this type of dental stress cannot be treated by currently known stress reduction techniques.

7.2 Dental Distress Syndrome is More Than Just Discomfort from a Problem Tooth

It may surprise many people that discomfort while eating is just the tip of the iceberg, one of many problems caused by Dental Distress Syndrome. The Autonomic Nervous System consists of two parts, the Sympathetic Nervous System (SNS) and the Parasympathetic Nervous System (PNS). Both of these branches innervate the mouth and coordinate the complex signals and sequence of events that occur in the organs of the body, and allow the organs to work together. Over activation of the SNS and under activation of the PNS is an underlying cause of all disease and any alteration or imbalance in these systems will interfere with the body's overall performance. [83]

Our modern understanding of Dental Distress Syndrome is reaffirmed in the principles of Traditional Chinese Medicine; although traditional Chinese practitioners did not have knowledge of modern neurology, they still arrived at similar conclusions. The *qi* meridians circulating throughout the body intersect at the mouth. Moreover, the diagrams illustrating the acupuncture points and meridians show the correspondence between each tooth and a specific organ and/or part of the body. It follows then that any change in the natural configuration of the teeth and the entire oral cavity can affect the general flow of *qi* and ultimately cause disease.

Those rare individuals who practice both dentistry and Traditional Chinese Medicine would agree that any alteration of a patient's teeth affects the sympathetic and parasympathetic nervous systems and the meridian system. It is extremely unfortunate that most dentists have little exposure, let alone formal training, in the Chinese medical art of acupuncture. It is equally unfortunate that most dentists do not understand dental neurology and Dental Distress Syndrome. The Dental Colleges do not train new practitioners to appreciate the effect of dentistry on the entire body (holistic dentistry), and the importance of the oral cavity as a vital nerve center and avenue for the flow of *qi.*

During a conversation with Naturopath, Ralph Weiss, ND, he emphasized the systemic importance of the oral cavity. He related that the director the Oregon State Mental Institution estimated that ninety percent of all people in the mental institution had lost the first and second molars on at least one side of their mouth. The anatomical components of the oral cavity are indeed closely linked to the body's energetic life force and state of health.

7.3   Neural and Dental Components Originate from the Same Embryonic Cells

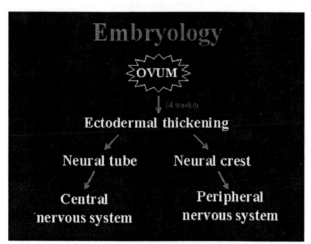

Fig. 27

A close association between the mouth and the nervous system can be traced back to the growth of a baby in its mother's womb. Some basic concepts in embryology and the functioning of the Autonomic Nervous System provide answers to the occurrence of Dental Distress Syndrome. The most complex structure within the human embryo, the human nervous system, is the earliest organ to develop and the last to be completed. The teeth, which affect the entire body, are present early on. To explain, let's review some basic Embryology (Figure 27). The egg and sperm unite and form a fertilized zygote. At four weeks an ectodermal thickening starts, which develops into the neural tube and the neural crest. The neural tube develops into the Central Nervous System and the neural crest develops into the Peripheral Nervous System.

After the egg and sperm unite to form a fertilized ovum, the zygote, the embryo becomes attached to the wall of the uterus. By the seventeenth day of development, the embryo has formed into a flat, elongated shape. The neural plate, a bulge in the middle of the embryo

Neural Tube	*Neural Crest*
Brain & Spinal cord	All sensory Receptors
CNS	PNS
½ of master pituitary	½ of master pituitary
Thalamus & hypothalamus	All other hormonal glands
Outgrowths of forebrain	
Midnose	**Balance of the**
Upper lip	**dental system,**
Premaxilla	**except tooth enamel**
Four maxillary incisors	

Figure 28

on the side facing away from the lining of the uterus, contains the cells that will become the nervous system. When this structure reaches its maximum thickness, it begins to enfold, allowing a hollow space shaped like a groove to form. Just before the folds on each side of the groove join to form the neural tube, neural crest cells near the edges of the folds move out of the way and an ectodermal thickening appears. This separates to form the neural tube and the neural crest.

The neural tube develops into the Central Nervous System (CNS), which includes the brain, spinal cord, thalamus, hypothalamus, mid-nose (also known as facial bone), upper lip, pre-maxilla (upper jaw), part of the forebrain, half of the master pituitary gland and the four maxillary central incisors (the front teeth). The neural crest forms the peripheral nervous system, all the sensory receptors, the other half of the pituitary gland, all other hormonal glands, and the rest of the dental system, except for the tooth enamel. (Figure 28)

The neural crest cells function to gather sensory information for the Central Nervous System, and the over-all body, through two processes. The first, called "proprioception" is an automatic sensitivity mechanism in the body that sends messages through the Central Nervous System and the Peripheral Nervous System. These systems then relay information to the rest of the body about how, and with what amount of tension, to react. The second process "kinesthetics," is the ability to know where your body parts are in three-dimensional space; this is required for every move we make.

These processes are performed by the Peripheral Nervous System, which originates from the brain and the spine and innervates the rest of the body. It is referred to as the Autonomic Nervous System and consists of the Sympathetic Nervous System (SNS) and the Parasympathetic Nervous System (PNS). When the information coming into the Central Nervous System is faulty, the outbound instructions are also inaccurate. The brain functions like any computer, *i.e.*, "garbage in—garbage out." [84]

The importance of the dental area to the entire body is illustrated in Figure 29. The Mandibular Nerve, also known as the Trigeminal or 5th Cranial Nerve, has fibers connecting directly to the Vagus Nerve. The Vagus Nerve not only innervates and controls the heart and circulatory system, but also controls the entire endocrine-organ system including the digestive, respiratory, reproductive, and elimination systems.

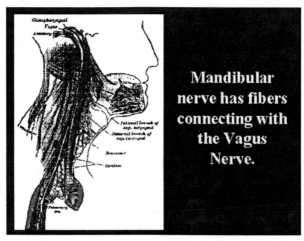

Fig. 29

Most dentists have not been trained in this delicate neurological balance and therefore do not understand the mouth's affect on the rest of the body. Therefore, before discussing Dental Distress Syndrome, proprioception and kinesthetics, a brief review of some basics of neurology is in order. Dentists have an awesome responsibility; what they do or don't do in the oral cavity can affect the body in so many ways. A dentist really should be a Neurologist of the highest order, but unfortunately this critical information is not included in the dental curriculum.

7.4 A Review of the Two Parts of the Autonomic Nervous System: the Sympathetic Nervous System and the Parasympathetic Nervous System

Most health care providers would agree that balance of the Sympathetic Nervous System and Parasympathetic Nervous System results in good health! A brief review of the Autonomic Nervous System will help you better understand this important balance and its overall effect on health.

The Autonomic Nervous System regulates vital functions of the body that are almost always involuntary (unconscious); these include all of the activities of the heart, organs and glands as well as the smooth muscles in the artery walls and in the colon.

Within the Autonomic Nervous System, the Sympathetic Nervous System:

- speeds up heart rate
- narrows blood vessels
- raises blood pressure

The Parasympathetic Nervous System:

- slows the heart rate
- increases intestinal and glandular activity
- relaxes ring-like muscles (sphincters), which close passageways and opens arteries

The Autonomic Nervous System —consisting of both the Sympathetic Nervous System and the Parasympathetic Nervous System —involves multi-millions of fiber bundles that connect the brain and the spinal cord with other parts of the body. These fibers carry inward-moving (afferent) signals from receiving organs toward the brain and spinal cord, and outward-directed (efferent) signals from the Central Nervous System to the various organs and tissues. This intricate signal-carrying set of nerve fibers activates, oversees, and controls all of the bodily functions.

There are specialized sensory nerve endings that monitor internal changes in the body brought about by movement and muscular activity. Proprioceptors are located in muscles and tendons and transmit information that is used to coordinate muscle activity.

The Autonomic Nervous System regulates these organ functions by coordinating the sympathetic and parasympathetic signals. When the Sympathetic Nervous System is stimulated, there will be an increase in body activity, stress, blood pressure, and heart and breathing rates. While these areas are activated, there is a simultaneous decrease in glandular, stomach and intestinal function. The body becomes more acidic, and has a greater tendency to stress and disease. Just the opposite occurs if the Parasympathetic Nervous System is stimulated: the heart and breathing rates slow down and the blood pressure and acid levels normalize. There is an increase in the glandular and gut activity. Body reserves increase and, generally, there is less disease.

Therefore, neurological specialists and other health professionals are correct when they say that balancing the Sympathetic Nervous System and Parasympathetic Nervous System will result in the occurrence of less disease. But they differ in their approach to achieving this balance: orthodox (allopathic) medical doctors may prescribe drugs, counseling and life style adjustments such as changes in the workplace and relationships, and taking more vacations; complementary and alternative (holistic) health care providers may agree with the lifestyle adjustments but also recommend exercise, dietary changes, supplementation, homeopathic remedies, chiropractic care and other alternatives. One important factor to keep in mind is that the sympathetic/parasympathetic balance is multi-faceted, and the factors that lead to balance or imbalance can and will vary from person to person.

Most health practitioners miss a crucial feature of the sympathetic / parasympathetic balance equation: the mechanics of the skull and jawbones. Many serious problems can be caused by a narrowing of the skull and upper arch, and/or a trapped mandible. If you lose some or all of your back teeth (not counting the wisdom teeth or 3rd molars), or if you lose the height of these back teeth (through natural wear and excessive grinding from an accident or because they have been pulled by your dentist), then the front teeth will meet before the back teeth. Remember, in the above discussion of embryology (Figure 28), it was mentioned that the four upper front teeth are extensions of the brain and spinal cord and have a major stimulatory effect on the Sympathetic Nervous System. It follows then, that there are serious repercussions when the relationship of the lower jaw to the skull is out of balance and the front teeth hit together too soon—the Sympathetic Nervous System is activated and the Parasympathetic Nervous System is deactivated. This causes chronically tight dental muscles leading to faulty proprioception to the brain, which in turn can lead to serious health problems. If your front teeth hit together too soon, you may come to have all the symptoms of stress, without the "normal" causes of stress (lack of food, shelter or safety, or financial insecurity, sickness or death, etc.) being present.

To reiterate, the increase in the activity of the Sympathetic Nervous System and a corresponding deactivation of the Parasympathetic Nervous System that results from the front teeth hitting together too soon can result in the following:

- Increased body activity
- Increased stress
- Increased blood pressure
- Increased heart and breathing rate.
- Increased body acidity
- Decrease of glandular, stomach and intestinal functions.
- General increase in stress and occurrence of disease.

If the missing back teeth are replaced, or if the height of the existing back teeth is corrected in accordance with the proper X-Y plane, (especially if done in conjunction with a program of low level laser therapy), then there is a good chance that the Sympathetic and Parasympathetic Nervous Systems will return to balance and the disease process will diminish. (Please see, *The Dental Physician* by Fonder.[1])

7.5   The Parasympathetic Nervous System

Like the conductor of an orchestra, the Parasympathetic Nervous System coordinates the thousands of chemical processes that occur simultaneously in the body. The Parasympathetic Nervous System, the controller of the body's organ systems, signals organs to start up and work in concert to accomplish a common goal. The complex daily coordination of the digestive system by the Parasympathetic Nervous System illustrates this process.

7.6   Digestion as a Parasympathetic Nervous System Process

Digestion, beginning with the smell and sight of food, is followed by the physical ingestion of it in the mouth. The front teeth tear or incise, and the back teeth grind, the food. Earlier we mentioned that all the teeth, except the upper and lower front teeth, are linked to the Parasympathetic Nervous System in the embryonic development of the nervous system. The process of food grinding, therefore, stimulates the Parasympathetic Nervous System. This is also why we desire a peaceful time for our meals—the function of digestion performs better in a Parasympathetic Nervous System-dominant state. The more we chew, the calmer we get. In a stressful situation, blood flows to the peripheral tissues for a possible dash for life. There is no thought for a joyful sit-down meal.

Much of the digestive process occurs in the stomach under the direction of the Parasympathetic Nervous System. Four or five times every minute, rippling waves controlled by the Autonomic Nervous System pass through the muscles of the stomach walls, mixing food with gastric acid and digestive enzymes secreted by the glands in the wall of the stomach. The food is reduced to a thin liquid mass and the digestive juices break down some of the nutrients into forms that can be utilized by the body. Carbohydrates, initially acted upon by the saliva during the chewing process in the mouth, continue to be broken down in the stomach; gastric juices and enzymes also begin breaking down protein and fat into forms that can be absorbed.

Peristaltic action continues to move the partially digested food through the stomach where the liquid food mass exits through the muscular pyloric valve to the duodenum, the initial portion of the small intestine. The emptying of the stomach is a complex act requiring an integration of feedback messages carried by the vagus nerve to and from the brain, as well as by intestinal hormones released in accordance with changes in volume and acid levels.

As food enters the upper part of the small intestine, hormones, stimulated by its arrival, coordinate the flow of digestive enzymes from the pancreas, and bile from the liver and gallbladder. A healthy pancreas that receives proper nerve stimulation via the Parasympathetic Nervous System produces sodium bicarbonate which neutralizes the acid contents of the stomach and gives the duodenum and intestines their alkaline environment, in contrast to the acidic environment of the stomach. Failure of the pancreas to properly function is one of the major causes of over acidic bodies. The pancreatic digestive secretions complete the breakdown of proteins and fats, processes that take longer than the digestion and conversion to glucose of carbohydrates. The small intestine is about 22 feet long, and as the now liquefied and well-mixed food continues its journey, a steady flow of basic nutrients becomes available to the body.

The small intestine is lined with villi, fingerlike projections that greatly increase the total surface area of the intestines. The villi permit each cell to come in contact with a one cell thick, microscopic blood vessel called a capillary, which permits the direct exchange of biological chemicals between the intestine and the blood. (A similar physiological mechanism to increase the surface area for blood-organ interchange occurs in the lungs and kidneys.)

Digested nutrients, sugars (from carbohydrates and some protein) and amino acids (from protein) pass through the villi directly into the bloodstream, and are carried to the liver or muscles to be further metabolized as ATP our body's fuel. One of the benefits of low level laser is that it enhances cellular production of ATP by as much as 150%. The digested fats pass from the villi to the lymphatic system or enter the bloodstream through the intestinal-hepatic portal vein. Blood, rich in digested nutrients, flows from the small intestine into the liver, which, acting in concert with the endocrine and other body systems, regulates the amount of nutrients (particularly glucose) that enter the bloodstream for distribution to various body tissues.

The Parasympathetic Nervous System signals the stomach and intestines to increase the flow of digestive juices and motility so that the digestive process continues without disruption. The Vagus nerve signals the heart to increase blood flow through our digestive organs and throughout the body. Each cell comes in direct contact with a blood capillary, which brings it oxygen and nutrients from the digestive organs and removes cellular waste.

The liver, the largest of the internal organs, accomplishes a number of highly complicated chemical processes. It produces bile, essential for the digestion of fats. One of the breakdown products of hemoglobin gives bile, and hence the stool, its brown color. The liver manufactures a number of other substances, including cholesterol, enzymes, vitamin A, blood coagulation factors, and complex proteins, as well. The liver also acts as a storehouse for blood, certain vitamins and minerals, and fuel, in the form of glycogen, which is readily converted to glucose as the body needs it. The liver, the major detoxifying organ, detoxifies prescription drugs, street drugs, alcohol and many other potentially harmful chemicals.

Finally, food that is not utilized by the body moves from the small intestine into the large intestine, the colon, where water is extracted from the waste material. The remaining waste moves, via peristalsis, through the large intestine to the rectum for eventual elimination in the form of a bowel movement. In order to maintain a healthy non-toxic body peristaltic action must function properly. Peristaltic action is controlled by the Parasympathetic Nervous System which gets its signals from the back teeth. If you are missing your back teeth or your posterior teeth are too short, peristaltic action will be impaired. Just as at the beginning of the digestive process, where the selection and placing of food in our mouth is initiated by our conscious control, elimination of waste products from the body at the end of the digestive process is also under our conscious control.

7.7   The Importance of a Healthy Parasympathetic Nervous System

Few of us truly appreciate just how precise and fine-tuned the body is—until something goes awry. For example, if the above-described digestive processes were not completely coordinated, no matter how much good food a person consumes, the body will not have sufficient raw materials. Our organs are interdependent; the smooth functioning of each organ depends upon the proper functioning of the others. The appropriate timing, necessary for overall coordination, is maintained by signals from the Parasympathetic Nervous System.

A life style that provides one's body with nutritious food, adequate sleep, daily exercise, mental stimulation, and the proper amount of stress will contribute significantly to good health for a long time. But despite one's vigilance and best efforts, a person can still suffer from chronic health problems if the Parasympathetic Nervous System is impaired. If suffering from an illness of undetermined origins, it is important to consider that symptoms affecting one part of the body may have their origin in a distant, and seemingly unrelated, organ. In many cases, the hidden problem lies in the oral cavity. Dental distress, often caused by missing back teeth or back teeth that are too short, leads to a condition of dysfunctional proprioception. This, in turn, puts the Sympathetic and Parasympathetic Nervous Systems out of balance and the end result is stress. The fact that the Parasympathetic Nervous System is not able to coordinate its signals to the internal organs properly, can lead to systemic health problems. For good health it is necessary to maintain a balance between the Parasympathetic and Sympathetic Nervous Systems.

7.8 Facial Muscles Illustrate the Important Connection between the Mouth and the Nervous System

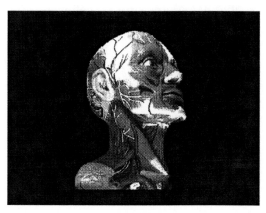

Fig. 29

The close relationship of the mouth and the nervous system can be illustrated from a perspective other than the digestive tract, through observation of emotional expression of the facial muscles. There are 68 pairs of muscles (Figure 29) above and below the mandible. These muscle groups control the posture and movement of the uppermost part of a person's body in the head, jaw and shoulder area. When overworked and tight, these facial muscles bring on faulty proprioception to the brain, which in turn affects the entire body. Every emotion has its own expressive facial movement. Silent film star, Charlie Chaplin, demonstrated how much can be communicated through the subtle nuances of our faces. This is possible because the human face has so many muscle functions:

- The muscle of the cheek, used for whistling and sucking, is involved in chewing.
- Another facial muscle forms a sphincter around the mouth; it purses the lips (as in kissing) and is also used for whistling and sucking.
- Another muscle raises the corner of the mouth. Its co-worker muscle, extending from the corner of the mouth towards the ear, is involved in grinning.
- A muscle located over the cheekbone is involved in smiling and laughing.
- Then we have a muscle which elevates the upper lip (as in sneering when we show contempt) and deepens the nasolabial sulcus (the rim around the upper mouth that gets deeper and more prominent with age) to express sadness.

- One of the dental muscles in the face curls the upper lip inside out and aids in expressing sadness.
- Another situated on the chin, pulls down the lower lip when expressing impatience and the counter muscle, elevates and puckers the skin around the chin and protrudes the lower lip when we express doubt.
- One muscle, located just under the nose, flares the nostrils when we are frightened or angry.
- The small muscle running from the bridge of the nose to the forehead is used in frowning and in reducing the glare of bright lights.
- Still another of the 68 pair of dental muscles elevates the eyebrows when we frown and express surprise.
- Many of these muscles are in bilateral pairs, one on each side of the face.

There are also muscles below the mouth and chin that extend along the neck to the shoulders and chest. For example, when we depress a sheet-like muscle covering the lower face and neck, it presses and wrinkles the skin allowing us to express displeasure. It is also used to depress the jaw (mandible), which opens the mouth.

Facial expressions seem to occur effortlessly. In reality, they are the result of a very complex neurological process that signals the muscles to respond. For example, the *platysma*, one of a pair of wide muscles at the side of the neck controlled by the facial nerves, develops in the embryo in conjunction with major components of the nervous system. When the platysma fully pulls back, the skin over the collarbone is drawn toward the lower jaw, making the neck appear larger and the person more formidable.

There is a vast network of nerves connecting these muscles to the Central Nervous System and it is these "dental muscles" that send signals to the brain via proprioception (part of the controlling mechanism of the Autonomic Nervous System). Like the process of digestion described previously, these muscles must work together in harmony in order to have a healthy well balanced body.

7.9   The Position of the Jaw Affects the Entire Nervous System

The Academy of Physiological Dentistry did research on the function of the mandible (lower jaw) from an engineering point of view and determined that the physical relationship of the lower jaw to the skull affects the entire body. Casey Guzay, an engineer and dental laboratory technician, along with other pioneering members of the Academy, developed the "Quadrant Theorem." This valuable research proved that the jaw does not articulate in the Tempromandibular Joint (TMJ) as most dentists are taught. Rather, the jaw's actual center of rotation is a point between the first and second cervical vertebra, specifically at the *dens,* and the jaw articulates from the scull by means of 68 pairs of muscles. Any imbalance in this delicate structure puts excessive demands on the muscles; their need to continually adjust so as to maintain an acquired "bite" or "occlusion" seriously impairs the signal system and throws the Sympathetic/Parasympathetic Nervous System out of balance.

This theory was studied and reaffirmed in an extensive clinical study of over 50,000 or more subjects by Dr. Koichi Miura of Japan.[85] Dr Miura first studied guinea pigs, then beagle dogs, monkeys and finally humans.   His monkey studies forever changed my mind about the importance of the relationship of the lower jaw and the skull. Healthy two-year-old male monkeys were put to sleep and impressions taken of both the upper and lower teeth. Then the teeth on ONLY ONE side were ground down, not enough to cause pain, but just enough so that they were "out of occlusion." The monkeys' behavior was observed and blood studies were taken regularly. There were immediate changes. They were unable to swing from trees with the arm on the side where the teeth were ground down. They became antisocial, fought with their siblings and did not eat well. The blood studies showed that several blood parameters had changed.

These results inspired Dr. Miura to begin human studies. First he ran blood studies and nearly 100 other medical parameters. After dental impressions were taken, a 10 mm-high flat plane template (also called a splint) was made to be worn at night, and a 5 mm-high flat plane template was made to be worn during the daytime and when eating. For one month, one or the other of the splints was worn at all times, except when cleaning the teeth. Then all the tests were re-run. There was a 75% improvement in approximately 75 of the 100 parameters run at the end of one month. Basically, this study demonstrated, using medically accepted diagnostic tests, that diseases were reversed after one month of template therapy.

These results should cause us great concern. The results of blood tests are routinely used by doctors to prescribe drugs or perform surgery for serious disorders such as cancer, heart disease and diabetes. This large study has yet to be duplicated. However, when it is combined with studies done at the Harvard School of Public Health and the hundreds of thousands of clinical case histories demonstrating the importance of the relationship of the jaw to the skull in balancing the Autonomic Nervous System, this important link should not be ignored. In fact it should be the first thing any health practitioner checks when coaching his patients back to health.

The head of the jawbone, called the "condyle," fits onto the skull in a loose arrangement that is held in place by 68 pair of muscles. Between the head of the condyle and the fossa in the skull is a fibrous soft tissue called the "disc" or "meniscus." The clicking which is often heard in the jaw is the condyle popping off the edge of this disc. The Tempromandibular Joint (TMJ) is not a true ball-joint configuration like the thighbone (tibia) that fits into the hipbone socket. The TMJ, which is moved and shifted in all directions by the muscles, is not restricted as are other joints in the body. That is why the skeletal remains of prehistoric humans are often found with the skull, but missing the lower jaw. It is not because the jaw broke off; instead, it was only attached to the skull by soft tissue which deteriorates quickly after death.

Because it is attached by this flexible soft tissue, the jaw is able to move in a variety of directions: up and down to open and close the mouth; side to side and forward and back to bring the teeth together in a grinding motion. The TMJ is flexible enough to allow the mouth to be opened very wide, changing its overall shape to accommodate something very large.

The spine or backbone is located just behind the jaw; the mandible and the spine are perpendicular to each other at about a right angle when the mouth is closed. As mentioned previously, the Quadrant Theorem demonstrates how the jaw articulates at the dens between the first and second cervical vertebra by a complex group of muscles. When the jaw is out of alignment physical stress is put on the spine causing neck, shoulder and back pain (Figure 30).

FUNCTION FROM A CENTER IN SUB-ATLAS AREA

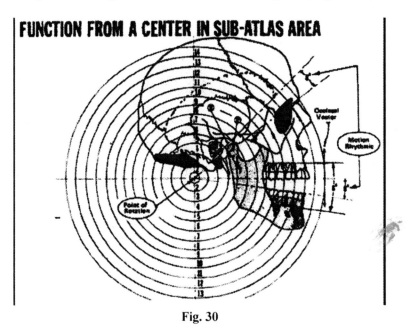

Fig. 30

A combination of micro-forces is exerted on the backbone by the jaw, which is controlled by powerful muscles. Engineers can calculate the centrifugal force and angular momentum when the jaw is moving down and up to open and close the mouth. The downward movement is augmented by gravity. Even when the mouth is closed, it is held in that position by strong muscles that keep the jaw in place.

These micro-forces from the jaw area are exerted backwards towards the spine. Like all energy, the force that is transmitted to the spine invades all the nearby structures as well. The Central Nervous System is certainly affected because the jaw is adjacent to the upper part of the backbone where the most crucial area of the spinal cord and the neural network is located. This is the part of the nervous system that controls and transmits many Parasympathetic Nervous System and Sympathetic Nervous System functions. Any disturbance or injury in this area can severely affect the function of the entire body. If a person sustains a traumatic back injury in this area, they can be rendered helpless, without any ability to move the limbs or torso. A completely severed spinal cord at the fifth vertebrate or above will cause death due to the total breakdown of the Autonomic Nervous System.

The anatomical proximity of the mouth and jaw to the upper part of the spine explains why the dental structures and the Central Nervous System develop in the embryo from the same tissue. In other words, the teeth (all except the enamel) are extensions of the brain and spinal cord. One great researcher, Nobel Laureate Tinbergen said: "a dentist should be a Neurologist of the highest order!" Unfortunately the dental schools do not teach and emphasis the importance of this critical aspect of health. The feedback mechanisms from the mouth to the brain that are paramount in controlling the Central Nervous System as well as the Autonomic Nervous System are known as mechanoreceptors and nociceptors. In this book they are referred to as "proprioception."

7.10   General Characteristics of Proprioception

People often wonder where the brain gets its information. Some speculate that it would take a gymnasium full of modern computers to do what the brain does. The fact is that the brain derives much of its feedback information from a process called "proprioception." What does this word mean? Why is knowledge of its definition important to your health and well being? Whether you are a health professional or a medical consumer, it is important that you understand this term. It may be an important factor in helping you restore and maintain your good health.

Proprioception, as stated above, is defined as stimulation of the body tissue to activate protective mechanisms. Interestingly, the proprioception signal to the brain does not travel by nerves—it is more like a sixth sense. The body responds through its voluntary nervous/muscular system, but in an involuntary way.

The Central Nervous System is a network of nerve fibers that extend everywhere throughout the human body. These nerve fibers send signals to the organs and muscles, and the nerves and muscles send signals back to the brain. All of this signaling occurs automatically and is not under the individual's conscious control. This process, called proprioception, is best understood and illustrated in the context of sports.

We can position ourselves to make the winning catch as we're running down the football field, but it's our Central Nervous System that actually makes our hands reach out, open, and get ready to grasp the ball. The moment our fingers feel the ball, a signal is sent to the brain. The brain then almost instantly relays a message to tighten our hand muscles around the ball. This "flash" of signals to and from the brain, necessary for the success of any ball player, is proprioception.

Much is known about proprioception in the feet and hands. For instance, if you were barefoot and stepped on broken glass, you would immediately (without thinking), lift your foot. The muscles in one leg tighten to lift the foot and the muscles in the other leg splint to support your weight. If this were to continue every step, your legs would quickly tire and you might think, how stupid of me to be walking barefoot when I just saw someone drop and break a glass? I should have put on my shoes.

Being warned in this way by proprioception, it doesn't matter whether you put on wooden shoes or leather shoes. Once you do put on shoes, your brain knows at once that you can now walk on broken glass. Another example of proprioception occurs if you were to touch an electrical wire: it shocks you instantly, and without thinking you jerk your hand away. This illustrates how the normally voluntary muscles can be controlled by in-voluntary proprioception signals.

However, proprioception is not only in operation in the hands and feet. A proprioceptive system, more delicate than anywhere else in the body, exists in the relationship between the upper and lower front teeth. Proprioception in this area is so delicate that one can tell the difference in thickness between an eyelash and an eyebrow. It is little wonder that it has such an effect on the Autonomic Nervous System.

7.11   Dental Proprioception

As mentioned earlier, the lower jaw is attached to the skull by 68 pair of muscles and stress on these muscles sends a feedback mechanism to the brain referred to as "dental proprioception." When these muscles are required to work "overtime" to keep the jaw in a position relative to an underdeveloped skull, they tighten, become tense and send false signals to the brain. These signals are sent to the thalamus which in turn controls the hypothalamus. The former controls the cerebellum and posture, and the latter controls the stress mechanism. Newton's Third Law of Motion states: that for every action, there is an equal and opposite reaction. The cerebellum requires proper information from the 5th cranial (mandibular) nerve to proceed with postural adjustment. Thus even small distortions in the dental proprioception easily reflect in the neck, shoulders, arms, lower back, legs and feet.

Drs. Penfield and Rasmussen, Neurologists and authors of *The Cerebral Cortex of Man,* state that almost half of the sensory and motor components of the brain are devoted to the "dental area."[86] The relationship of the lower jaw to the skull affects over 50% of the bodily functions including, motor and sensory actions, and blood supply to the brain. Note that I did not say "bite" or "occlusion," nor did I say TMJ. Everyone has a bite and a TM joint and even if the bite appears normal and there is no evidence of TMJ disease, they could still have faulty proprioceptive feedback to the brain through the 68 pair of dental muscles that connect the jaw to the skull.

Drs. Weston Price and F.M. Pottenger in the classic book, *Nutrition and Physical Degeneration*, states that the skull, particularly the maxilla and premaxilla, are getting narrower in each succeeding generation in populations that consume mainly refined food diets.[87] This observation is supported by Dr. Francis Pottenger's cat studies.[88] This narrow underdeveloped upper arch influences the 68 pairs of muscles that regulate the mandibular position, which causes the firing of faulty signals to the brain. These 136 muscles are supposed to be reciprocally balanced, and able to contract and relax naturally at their respective lengths. When they are unable to do so, and become tight, then faulty proprioceptive feedback occurs. When these "dental muscles" are not in a state of homeostasis, dental distress syndrome is perpetuated, resulting in tight muscles, which in turn cause a reduced blood flow to the brain and all the associated structures of the head. It is believed that this structural problem is a factor in most diseases. As the large 1986 Japanese study shows (see 7.9 above), placing a template or splint to increase the height of the back teeth and activate the parasympathetic nervous system improved 75 out of 100 medical parameters including blood cholesterol and LDL levels.[89]

As mentioned earlier in this chapter, dental distress is more than just a tooth problem. Because of the oral cavity's delicate relationship to the brain, any dental problem that exists can easily affect the entire sympathetic/parasympathetic nervous system. Any injury or medical condition can cause nearby nerves to malfunction, and send incorrect and unhealthy signals to the brain. It is well known in the rehabilitative medicine community, that, "the body follows the head." If the posture of the head, neck and shoulders can be corrected, the rest of the body will fall into line. The same axiom comes into play when treating the unhealthy stimulus caused by faulty dental proprioception.

As aware as we are of the many types of proprioception, nearly all people, including many health professionals, are unaware that the most delicate proprioception in the entire body, including that within the eyes, ears, fingers and feet, actually lies inside the mouth–specifically between the upper and lower front teeth. The proprioception here is so delicate that you can tell when any object as thin as $1/1,000,000^{th}$ of an inch is placed between your front teeth! Elimination of faulty dental proprioception will benefit your health and well being more than any other single thing you can do.

At one time, health scientists assumed that this delicate proprioception mechanism resided in each and every ligament that anchors the teeth into the bone. But what happens if you loose a tooth? Do you loose proprioception? Do you lose proprioception if you were to get dental implants? If you loose all your teeth and must wear a denture, is proprioception lost? The answer to all of these questions is NO. Proprioception remains in the environment where the body needs it to be for protection. But there is good proprioception and faulty proprioception.

7.12 Faulty Proprioception

What happens when one's proprioception is faulty, when it gives false signals that do not benefit the body? Dysfunctional proprioception causes harm because it sends signals that bring about a Sympathetic/Parasympathetic Nervous System imbalance. Most often this faulty proprioception originates within the belly of one or more pairs of the 68 muscles that are involved in positioning the mandible to the maxilla. For simplicity's sake, these 68 pairs of muscles in the face will be referred to as "dental muscles."

To understand why some of these dental muscles work overtime to control the jaw position, it is necessary to review the development of the skull. When an egg and sperm meet, the brain, spinal cord, endocrine glands, nervous system and all of the teeth begin to form. Yes, as stated above, the teeth are extensions of your brain and spinal cord. Except for the dental enamel, they come from neurological tissue. Moreover, the development of the skull, and these other delicate structures, is directly related to the diet of the newborn's father and mother. Diets that include refined foods have been shown to cause a narrowing of the skull, brain cavity, and maxilla. This process continues in time. With each generation that lives on refined foods, there is a noticeable narrowing of the premaxilla, which forms the base of the nose and the upper four front teeth. It is this narrowing of the skull and underdevelopment of the premaxilla that causes most of the problems with faulty proprioception to the brain.

How does this narrowing process lead to faulty proprioception? When the mandible closes, the dental muscles must tighten up and move the mandible into a position dictated by the maxilla in order to have a so-called occlusion or "bite." This is referred to as "trapped mandible syndrome." Since the teeth come together once or twice a minute when swallowing, these dental muscles are always working. In fact, the dental muscles receive such a workout they easily become the strongest muscles in the body and are able to exert over 1200 lbs of pressure. They can move teeth, fracture teeth, crumble fillings and crowns, break metal bridges, and can even knock off a tooth at the gum line

The only time the dental muscles go into a state of homeostasis or relaxation is during speech. Observe an individual's mandible when they talk, and you can easily tell where the jaw needs to be in order to allow the muscles to relax. Most importantly, there is no faulty proprioception to the brain when the dental muscles are in a state of relaxation. All too often these powerful facial/dental muscles are in a state of tension and are sending faulty signals to the brain. It is so common in the United States; it could be considered an epidemic.

While all such action is destructive, few people realize that even more problematic and less recognized is the fact that these powerful muscles actually depress the back teeth and make them shorter. Technically, this is known in dentistry as loss of vertical dimension. As this process occurs, the nose and chin come closer together, one loses the full facial appearance and fullness of the lips, more wrinkles appear at the corners of the mouth and aging occurs more rapidly.

The most troublesome factor is that the loss of back teeth, or loss of the height of the back teeth, allows the front teeth to hit first when closing the mouth. The impact from the meeting of the front teeth over-stimulates the Sympathetic Nervous System. This, in turn leads to symptoms of stress, which may be misinterpreted as caused by something external such as personal problems, work related problems, noise, or traffic jams. An over active Sympathetic Nervous System affects the entire structural system and results in neck, shoulder, back, arm, hand, hip, leg, and foot pain.

The counterpart of the Sympathetic Nervous System, the Parasympathetic Nervous System, becomes under-activated when there is loss of posterior teeth or even the height of the back teeth. An improperly stimulated Parasympathetic Nervous System adversely affects the entire endocrine system and all of the viscera (internal organs). This chain of events is the result of faulty proprioception to the brain originating in the mouth.

A topic we have not addressed here is the bite (sometimes called occlusion) and Tempromandibular Joint (TMJ) problems. Even if you have, what might be called, a "normal" acquired bite and no problems with TMJ, you still very well might have faulty proprioception to the brain because the 68 pairs of muscles that control the position of the lower jaw have to work overtime to maintain the acquired bite or occlusion.

Proprioception affects sensory interpretations connected to vision, touch, taste, sound, and speech, and motor functions that deal with all bodily movements, including that of the hands, arms, shoulders, feet, legs, hips and even the length of the stride. All of these sensory and motor functions are controlled by proprioception to the brain that originates in the 68 pairs of dental muscles. But these muscles are also constantly called upon to correct the above-mentioned "trapped mandible" syndrome.

7.13   Drs. Weston A. Price and Francis Pottenger

"Why are these (dental) muscles so often out of balance in the mouths of people living in our modern world?" To get this answer we must refer to the pioneering research studies in nutrition conducted by Drs. Weston A. Price and Francis Pottenger. Dr. Price, a dentist turned medical anthropologist, found that when western refined foods were adopted by traditional societies, not only did these refined foods began to cause dental decay but they also caused a narrowing of the skull and upper dental arch which contributed to their general physical degeneration.[87] Francis Pottenger, M.D., a medical doctor, performed research on cats showing that those cats that had been fed refined or cooked food diets progressively developed narrowing skulls and narrowing upper dental arches when compared to cats that ingested raw food diets. Pottenger's experiments on cats are famous in the science of nutrition.[88]

The observations of Dr. Price, who studied many Pacific Island tribes and other isolated social groups closely resembled Pottenger's cat studies—refined food diets resulted in the narrowing of the skull and upper dental arch. When this landmark research is combined with the genetic changes in mixed races, there is a very noticeable and destructive narrowing of the skull and the maxillary arch, particularly a narrowing of the pre-maxilla.

This underdevelopment of the pre-maxilla leads to the condition called the "trapped mandible syndrome," which occurs when the jaw cannot functionally operate in the correct position. The improper positioning of the jaw causes the 68 pairs of dental muscles to work overtime to try to keep the jaw in its correct functioning position. Remember, this has nothing to do with "bite" or TMJ, regardless of which classification of occlusion (I, II or III) an individual might be classified. Regardless of whether they have "normal" or "malocclusion," there can still be faulty proprioception to the brain. A simple way to check for faulty proprioception is to palpitate or feel the muscles under the angle of the jaw or inside the mouth behind the upper 2nd molar. Any sore "dental muscles" will send faulty signals to the brain.

7.14   The Quadrant Theory and Dental Distress Syndrome

According to the Quadrant Theory (mentioned above), the easiest method for minimizing any unhealthy stimulation of the central nervous system by the relationship of the jaw to the skull is to provide proper molar support in the mouth—that is the molars have to be higher. It is crucial that upon normal closing of the mouth the back teeth hit first. The front teeth should then hit when the jaw is protruded forward or to the side. This is known as "anterior or cuspid guidance" by dentists. When the back teeth are too short or missing, the airway is narrowed and the breathing is affected; this reduces the amount of oxygen going to the brain.

The loss of posterior support leads to a "forward head position" and the poor posture that is seen in a large percentage of our population, especially the older people. The loss of posterior tooth height causes a serious health dilemma leading to less oxygen to the brain, the head comes more forward, the dental muscles become tighter, there are more faulty signals to the Autonomic Nervous System and the health in general continues to deteriorate. This dysfunctional dental proprioception is known as "Dental Distress Syndrome" and will continue to prevail and become more severe over time unless attended to.

The term, Dental Distress Syndrome was first coined by Dr. Alfred Fonder in his writings, *The Dental Physician*[1] and *Dental Distress Syndrome (DDS)*.[90] It is known that the thalamus regulates the hypothalamus, a part of the brain that is considered to be responsible for the stress mechanism. Dr. Hans Selye, the father of the modern theory of stress, states in the introduction to Fonder's book: "Stress, particularly stress of dental origin, pervades man's life in health and disease. Medicine would benefit by a closer alliance between members of the medical and dental professions."[81, 82] Unfortunately, Dr. Selye died before this alliance was achieved, and without his prominent influence it will take much more time to accomplish it.

Patients suffering from Dental Distress Syndrome experience a variety of health problems including: allergies, bladder and kidney complications, body pains and numbness, cold hands and feet, constipation, depression, dermatitis, dizziness, fatigue, forgetfulness, frequent urination, gynecological problems, headache, hearing loss, indigestion, insomnia, nervousness, sexual dysfunction, sinusitis, suicidal tendencies, ulcers, worry and a variety of other symptoms. Although the many new medical tests on the market may indicate a variety of causes for these conditions, actually, the ailments are often the result of a structural problem that only a handful of dentists are trained to treat. These chronic ailments are best alleviated by correcting the faulty dental proprioception problem.

7.15　Misdiagnosis of Dental Distress Syndrome as "Normal" Mental Stress

It is interesting to note that many of the symptoms of Dental Distress Syndrome are considered to be normal symptoms of stress. This may explain why Dental Distress Syndrome is often not properly diagnosed. After physically examining a patient and evaluating their test results, many doctors may conclude that there is nothing medically wrong. They may say: "It's all in your head," meaning that it is a psychological problem. Such statements do an injustice to the patients and it is hopeful that this book will enlighten doctors of all professions and help them learn more about the devastating effects of faulty dental proprioception.

In modern times, nearly all persons residing in western industrialized nations suffer from the symptoms of emotional and mental stress, but they are rarely confronted by the ever-present, life-threatening physical stressors that our forefathers had to contend with. Ideally, doctors would have the perspicacity to investigate causes of stress that might result from conditions other than life style. Unfortunately, too many physicians are unaware that faulty dental proprioception can cause Dental Distress Syndrome and that Dental Distress Syndrome leads to emotional and mental stress. The patient's discomfort, that often is assumed to be coming from a stress-ridden mental condition, is actually caused by an over-stimulated Sympathetic Nervous System and an under stimulated Parasympathetic Nervous System.　Evidence of this sort of misdiagnosis occurs when medical tests of body chemistry and blood return to normal values after dental proprioception is corrected.

7.16   Natural and Unnatural Dental Problems

Sometimes faulty dental proprioception can occur naturally from a mouth full of crooked teeth that do not line up properly; the back molars misalign the jaw with the rest of the head. Sometimes, the problem is caused by dental work. Every dental filling is placed just slightly lower than the previous filling that was removed. Extraction of back teeth can result in front teeth hitting to soon and over activation of the Sympathetic Nervous System ---a stress factor.

Removal of root canalled teeth, particularly back teeth, is not actually beneficial because the patient is left with faulty proprioception. All too often dental work that was intended to correct a problem in the patient's mouth now creates a condition that can affect the Central Nervous System and lead to systemic medical ailments.

Today we know that some dental extractions result in faulty dental proprioception and are downright dangerous to the patient's health. The human body has a tremendous ability to heal itself, but there are limits to its capacity to do this. Some species of animals can regenerate entire limbs, and others have generations of teeth that move into place when the more forward positioned teeth are lost. However, in the animal world when the teeth are lost the animal dies. Since humans cannot grow replacement teeth, whatever is done by a dentist is permanent. Based on our understanding of dental proprioception, all dental work can affect the function of the entire body.

It is ironic that many patients are currently suffering from faulty dental proprioception caused by dental work performed by professional dentists. This lack of understanding of dental neurology and dental proprioception on the part of dentists reflects a conflict between paradigms. But don't blame your dentist. Blame the dental institutions. They are not educating the dentists about dental neurology. As patients, we should view our teeth as a vital part of our general health and spend whatever amount of time is required to educate ourselves and in turn influence dentistry to change. There is a rather new specialty in dentistry, not recognized by the American Dental Association. The practitioners call themselves "biological", "holistic" or "environmental" dentists. These dentists are trained in the safe removal of mercury fillings and the use of compatible dental materials. They generally support a broader integrative role of dentistry. Unfortunately, even the holistic and environmental dentists are not adequately trained in neuromuscular dentistry. Because of this, they often fail to understand the role posterior support plays in balancing the Autonomic Nervous System. This situation leaves you, the patient, responsible for getting yourself informed and taking the responsibility for your own dental health.

7.17   Finding Ways to Balance the Autonomic Nervous System

Since there are so few dentists practicing neuromuscular/ proprioceptive dentistry, how can individuals help themselves balance the Autonomic Nervous System and reduce Dental Distress Syndrome? There is a way for you as a patient to take charge and that is by buying, making and learning to adjust your own proprioceptive guides.

Proprioceptive guides are devices that fit over the lower back teeth. They are temporary devices that provide a raised platform that places the jaw in proper alignment to the backbone. Increasing the height of the lower back teeth with the proprioceptive guides activates the Parasympathetic Nervous System and de-activates, or one might say, balances the Sympathetic Nervous System.

Proprioceptive guides come in two different types. Both MUST be adjusted by the patient in order to successfully correct the faulty proprioception. The first type, called Miracle Bite Tabs™ (MBTs) are designed to be used by people who have their lower back teeth, including the 1st and 2nd molars but excluding the wisdom teeth (Figure 31).

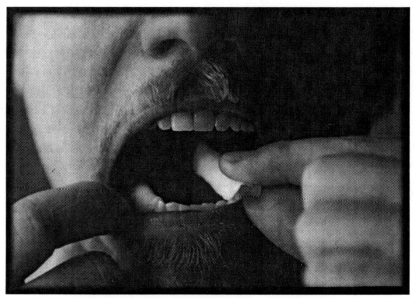

MIRACLE BITE TABS

Fig 31

Most Caucasians do not have room for their wisdom teeth and have had them removed.

MBTs can be made by lay people at home in their kitchens. The material is first softened with very hot water, then fitted over the lower back teeth and finally cooled with ice. MBTs feel like they're gently gripping the sides of the teeth. This allows the patient to wear the bite tabs while performing normal daily activities, such as talking, working and even jogging. The material is flexible so they can be removed for eating and cleaning by gently grasping them with the thumb or forefinger and carefully lifting them up.

The MBTs, as well as the Easy Adjust Proprioceptive Guides, are adjusted as needed by following the written or video instructions. Most people use a table knife dipped in very hot water. Some innovative and handy people even use a Dremel tool or miniature butane flame for making adjustments. MBTs will need adjusting as the dental muscles that control the relationship of the jaw to the skull change. Sometimes this happens in a few minutes and other times over a longer period of time. The material used for MBTs is inert and there have been no reported allergies to it.

If the lower back teeth are missing, the individual cannot use the MBTs and instead needs to use the second type of proprioceptive guide, the Easy Adjust Proprioceptive Guide (Figure 32). To use this guide, you must first have a dentist or assistant take impressions of your mouth and make dental models of your remaining teeth. These models are then sent to a special laboratory that will make the Easy Adjust Proprioceptive Guide and return it to you to be placed and adjusted as described. To obtain more detailed information about proprioceptive guides, please see the Resource section at the back of this book.

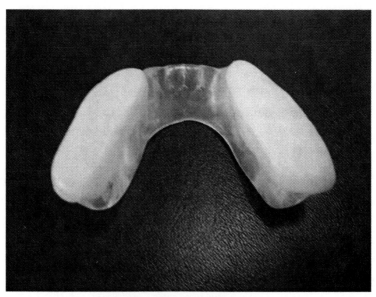

**EASY ADJUST
PROPROCEPTIVE GUIDE**

Fig 32

MBTs and Easy Adjust Proprioceptive Guides are not considered permanent and are used to temporarily reduce faulty dental proprioception to the brain and alleviate Dental Distress Syndrome. Some people find it easy to make and adjust the guides and take pride in having some control over their health. Others may not have the dexterity or desire to work in their own mouth and find the process more difficult. Those who find it difficult, need only find a "coach" to assist them. This could be a family member, a friend, or a professional.

When either the MBTs or the Easy Adjust Proprioceptive Guides are made and adjusted correctly, they are comfortable enough to be worn day and night except when eating.

Low level laser therapy and proprioceptive guides work synergistically to balance the Autonomic Nervous System. This means that when used together the benefits are greater than when used alone. The role of low level laser in correcting the symptoms of faulty proprioception is significant. Mode 1 of the Q1000 laser acts through biofeedback to reprogram the brain and balance the Autonomic Nervous System. For more information, we advise that you take two important actions:

(1) Go to the website: www.laserinformation.com and read Dr. Al Fonder's summary of dental distress syndrome
(2) Attend one of Dr. Larry Lytle's workshops posted on that same website.

The low level laser therapy protocol for faulty proprioception is as follows: apply mode 1 of the Q1000 laser under the angle of the jaw (internal ptyergoid) on the high shoulder side or the sore muscle side. Then laser over the TMJ, just in front of the ear over the lateral ptyergoid muscle. Third, laser the area just below the collarbone. Fourth, laser the shoulder blade on the high shoulder side. Fifth, laser the area in the groin where the leg meets the torso on the same side. And last laser the buttock on the outside edge of the pants pocket. The laser is used for 6-8 breathes on the side of the body that is sore, where the shoulder is higher or where the head is tipped indicating that the muscles are tight. This can also be determined by observing the high shoulder or on the side to which the head is tipped.

Remember all lasers are not designed to release faulty proprioception. Do not use higher power, single wavelength lasers on bellies of muscles. Too much power may produce no results at all or even make the problem worse. The Q1000 produces the correct power density and combination of wavelengths and frequency for quick muscular release. These lasers also have the patented soliton wave which allows the energy to penetrate deep into the muscle. Yet, unless you use Proprioceptive Guides (see GO-JO in appendix) or get an acceptable dental splint to use along with your low level laser therapy, the dental muscles will probably tighten and again produce faulty proprioception to the brain. Until you can either temporarily or permanently correct the lost posterior support of your teeth, regular home use of the laser is necessary.

Elimination of faulty dental proprioception with a proprioceptive guide or correct dental splint, and the use of the right low level lasers, in the appropriate order, and in the right places, can greatly enhance the well-being of humankind. This should be considered as a first line of treatment for any complaint.

Chapter 8   Applications of Low Level Laser Therapy for Other Dental Procedures

8.1   Introduction

There are more than 2,500 scientific studies showing the benefits of low level laser therapy, and more than 325 of those focus on dentistry. These studies were conducted in a total of 82 institutions in 37 countries. Although the quality of these studies varies, more than 90% of them reported positive effects of low level laser therapy on more than 24 different dental conditions. These include tooth hypersensitivity from crowns, fillings, exposed roots, root canal fillings, gums disease, Herpes simplex or cold sores, mucocities (a highly inflammatory condition associated with cancer treatment), post-operative pain from gum or dental surgery, parathesia of the dental nerve, sinusitis, TMD/TMJ, tinnitus/vertigo, trigeminal neuralgia, herpes zoster, bone regeneration, nausea from surgical procedures and other conditions.[91, 92, 93]

8.2   Low Level Laser Therapy for Tempromandibular Disorder (TMD/TMJ)

Since low level laser therapy has proven effective for many kinds of pain, it was decided to test it for TMD. In one study with 35 patients, low level laser therapy reduced pain and the number of tender points, and improved mouth opening and lateral movement of the mouth

Herbert Yolin, D.D.S., an enlightened general dentist practicing in Brookline, Massachusetts, with postgraduate certification in Prosthetic Dentistry from Boston University School of Graduate Dentistry, uses low level laser therapy for a number of dental procedures. Dr. Yolin is a member of the Academy of Laser Dentistry and has proficiency certification in the uses of lasers in dentistry. Dr. Yolin is also a member of the International Academy of Oral Medicine and Toxicology and has served on the staff of Boston University School of Graduate Dentistry. He has undergone extensive postgraduate training in TMD, cosmetic dentistry, implantology, and non-surgical therapy for periodontal disease. Dr. Yolin, who frequently lectures to other dentists, dental laboratory technicians and medical/dental consumer groups, testifies to the following:

One of my daughters had been suffering from an incredibly painful, stress related TMD problem. Despite all my dental knowledge and past experience in treating TMD, my help was fleeting and minimal. I lingered in a state of frustration! Moreover, my daughter was studying for her Masters degree as a Physician Assistant (PA) and was caught up in the western medical model. Alternative dental therapies were of no interest to her. Then she underwent a Magnetic Resonance Imaging (MRI) of her jaw, was diagnosed with TMD plus a dislocated jaw, and was advised to consider surgery and medication for their correction," explains Dr. Yolin. "I did not agree with such recommendations and emphatically told her she could only do that after I was dead and buried.

Dr. Yolin said:

My daughter's condition was so painful that she claimed to understand people ending their lives over that kind of pain. Time passed until I was finally able to convince her to try an alternative method of healing TMD: low level laser therapy in combination with a corrective proprioceptive dental device invented by Dr. Larry Lytle called "Miracle Bite Tabs." Well, the low level laser therapy plus Dr. Lytle's bite tabs worked beautifully for my daughter. Since her treatments she has been pain free!

Dr. Yolin relates an anecdote about a dental patient's head pain. He describes Betty as, a cooperative and friendly woman who called me one day looking for help to curb what was described as a 'stay in bed, about to throw up' migraine headache. What happened when she came for her appointment surprised even me. In a single two-minute demonstration using the Q1000 and 660 Enhancer lasers, I was able to eliminate my patient's migraine and all symptoms associated with it. Periodically, Betty consults me for laser treatment to eliminate migraine headaches which are now occurring much less frequently.

After hearing about Betty's experience, another woman in the audience volunteered her own story:

> I am Thulani DeMarsay of Brookline, Massachusetts. I had been experiencing severe tension headaches and TMJ pain, rendering me unable to drive or even do simple household chores. I underwent an MRI and CAT Scan with all tests coming back normal. The doctor prescribed pain medication which made me feel worse. One day, I learned about Dr. Herb Yolin and his holistic dental practice, and made an appointment. Dr. Yolin fitted me for a unique proprioceptive device called "Miracle Bite Tabs" and used acupuncture with his low level laser therapy at the site of acupoints on my jaw. He also applied the laser to my TMJ points. Within 10 minutes of receiving this treatment, I felt a release in my jaw and have not had any pain since. That pleasant experience occurred almost two years ago.

8.3   Low Level Laser Therapy for Dental Analgesia

Low level laser can also be used instead of needles and Novocain shots for dental analgesia, either by direct application of the laser to target tissue or via skin acupuncture points. In China, 562 cases of surgical exodontias and 48 cases of minor oral surgery were performed using low level laser on acupoints rather than needles. The doctors were able to provide effective pre and post-operative analgesia in all cases. Similar results were obtained in a separate study with 3,000 patients using low level laser directly on the dental sight. No sedatives or analgesics were given before or during any of the procedures.

Dr Bill Frost, an accredited laser dentist from Arizona, with a clinical practice in Pedodontics (children's dentistry), says he uses low level laser therapy in place of Novocain shots on approximately 75% of his active clientele. Not only does it save him a lot of time, but also both the children and the parents love the laser. It has helped build his practice.

The case studies show that, when applied as an analgesic, low level laser therapy decreases the firing frequency of nociceptors. Low level laser therapy selectively inhibits a range of nociceptive signals arising from peripheral nerves involved in various sensations, such as when the skin is pinched and is exposed to hot and cold temperatures and chemical irritation.

A patient explains her experience:

I am Susan Korn of Boston, Massachusetts. My dentist, Dr. Herb Yolin, first introduced me to low level laser therapy as an accompaniment to acupuncture. I had significant decay under an old filling and Dr. Yolin needed to drill close to the nerve to replace the filling. He told me that he had been using the low level laser for acupoint therapy, and that about 70% of the time, his patients did not need Novocain for my type of procedure. He asked if I was open to trying it, and I said yes! Dr. Yolin lasered the tooth at certain acupoints, and then started drilling. I heard the drill but felt no pain or discomfort! After he finished the filling he lasered the tooth again to reduce tooth sensitivity and speed the healing. I had absolutely no pain or discomfort during the process of drilling and filling my tooth without Novocain.

Another dentist, Dr. Martha Cortes of New York City, also uses low level laser therapy in her office, and agrees with Dr. Yolin on the merits of using laser for dental procedures. Many other dentists have bought Q Series laser devices and use them with great satisfaction.

8.4 Low Level Laser Therapy for Dental Nerve Regeneration

Low level laser therapy has been shown to promote axonal growth of damaged nerves in animal models. Thus, low level laser therapy may be ideal for dental nerve regeneration in humans as well. The incidence of damage to the Mandibular nerve during needle injection and removal of the third molars (wisdom tooth) is reported to be as high as 5.5%, while in sagittal split osteotomy, the incidence is up to 100%. There is some positive evidence that shows direct application of low level laser therapy helps nerve regeneration during these surgical procedures. Thirteen patients suffering from long-standing post-surgical nerve damage were treated with low level laser therapy and their mechanoreceptor sensory tests and subjective feeling improved. However, thermal response did not improve. In this study low level laser therapy only partially restored nervous functions.

8.5 Low Level Laser Therapy for Root Canal Teeth

A large number of in vitro studies have reported enhanced killing of bacteria using various dyes in combination with low level laser therapy. The most commonly used dye is toluidine blue (TBO). The bactericidal effect of toluidine blue against streptococcus mutants and staphylococcus aureus, two common species of oral bacteria, was increased by low level laser therapy.[43]

This finding has important implications for treating dental caries, periodontal disease and teeth with root canals. Dr. Weston Price found that many teeth with root canals were contaminated with infectious agents. Only 30% of human subjects appeared to remain healthy after undergoing root canal procedures. It is therefore imperative to find a way of removing the infections in root canals.

A recent in vitro study explored the possibility of reducing bacteria in teeth contaminated with *Enterococcus Faecalis* by using a dye and low level laser therapy. Thirty teeth with prepared root canals were contaminated with *E. faecalis*. Ten of those teeth received a sham chemical, sodium hypochlorite, for 30 minutes. Another group of ten teeth received the azulene dye paste for 5 minutes and were irradiated with a 685 nm diode laser outputting 10 mW for 3 minutes. The ten teeth in the control group did not receive treatment. The bacterial reduction was significantly higher for the group receiving laser, when compared to the group receiving the sham treatment and the control group. These results indicate that photodynamic therapy is an effective method for flushing bacteria out of the dentinal tubules.

My protocol for treating root canals is quite simple and can be used by either the dentist doing the root canal or the patient if the dentist does not have a laser. The root canal is prepared and sterilized according to good Endodontic technique. Just before the dentist places the root canal filling material, apply the 660 Enhancer laser for 30-45 seconds directly on the crown of the tooth. If there is bone damage at the tip of the tooth, apply the 808 Enhancer laser for 30-45 seconds through the skin over the tip of the tooth. For existing root canal filled teeth that are suspected of having bacteria in the dentinal tubules, use lasers in the same way. However, if the tooth is crowned with a metal crown, apply the laser at the gum line rather than on the metal crown.

Many cavitation surgeons and holistic dentists recommend extraction of teeth with root canals and removal of toxic filling materials. They believe that teeth with root canals should be extracted because they are "dead" teeth and harbor toxins. However, although the tooth has less vitality because it has lost its nerve and pulp, and the ability to feel heat, cold and sweets, nevertheless, a tooth with a properly treated and filled root canal still has a vital connection to the bone. The periodontal ligament, with its blood supply and nutrients, connects, via Sharpies fibers, to the cementum of the devitalized tooth. Therefore, technically, <u>a tooth with an existing root canal is not "dead."</u>

While these devitalized teeth are not dead, they do have bacteria in the dentinal tubules that produce infectious material that can spread to other parts of the body if untreated. This is referred to as a "foci of infection" and is the type of infection referred to by Dr. Price. While Dr. Price thought root canalled teeth should be pulled, today extraction is not necessary because they can be treated successfully with a laser.

Most often, the pain felt in an endodontically treated tooth is caused by an inflammation of the periodontal ligament that holds the tooth in the bone. While root canal pain can be caused by improperly filled canals, often it is due to heavy pressure caused by very tight muscles. These muscles get their signal to tighten from faulty proprioception to the brain and can put as much as 1200 lbs. of pressure on the teeth, thus creating the "sore tooth syndrome." Considering the importance of the back teeth in proprioception to the brain, every effort should be made to maintain healthy teeth, particularly the back teeth, even if this means having a properly treated and filled root canal. Despite what some of the naysayers of root canal therapy are publishing, it is possible to have a healthy root canal without it negatively affecting your health.

It is known that all organisms, including bacteria found in dentinal tubules, have a protective instinct to move away from something that threatens them. These bacteria do not like the laser light, and will move away from the light, out of the dentinal tubules and back into the blood stream where the body's immune system, homeopathic remedies or antibiotics kill bacteria. Because many root canalled teeth have lost their vitality due to trauma caused by strained muscles, it is very beneficial to make and wear MBTs when lasering troublesome teeth. Remember, the 68 pairs of muscles that control the lower jaw have over 1200 lbs. of power. The pressure applied by these muscles plays a big role in the death of the dental pulp. These muscles can be released by applying mode 1 of the Q1000 laser for six breathes (approximately 30-45 seconds) over each of the four Proprioceptive Points, which are located: (1) just in front of the ear over the TMJ, (2) under the angle of the jaw, (3) two inches below the collar bone, and (4) one inch up from the rounded angle of the shoulder blade (scapula). These proprioceptive points are pictured and described in my *Low Level Laser User's Manual*.[17]

The success of low level laser therapy for treatment of root canals can be measured in a number of ways:

- Reduction of pain and other symptoms.
- Electro-Dermal screening devices.
- Kinesiology performed by a qualified practitioner.
- Circles, a self method of Kinesiology described in *Understanding Low Level Laser Therapy* and the *Low Level Laser User's Manual.*
- Other energy techniques.

8.6   Low Level Laser Therapy for Dental Cavitations

Dental cavitations are known by many names, but to lay people, they are holes in the bone and should not be confused with cavities or holes in the teeth. These holes in the bone are technically referred to as Ischemic Osteonecrosis, Neurological Inducing Cavitational Osteonecrosis and Chronic Osteomyelitis, but are commonly called jaw cavitations or just plain cavitations. The history of cavitations dates back to the early 1900s; since that time they have been thought to be the cause of a variety of disorders because of the metastasis of the various microorganisms and toxins found in the cavitation sites. Current research, including that by Dr. Boyd Haley from the University of Kentucky, has documented the existence of various microorganisms in these jawbone holes. Opportunistic in seeking environments where they can flourish, healthy and unhealthy microorganisms, or bacteria, live everywhere in the body. Undoubtedly they can be found in holes in the jaw bone and other bones as well.

Various publications list the cause of cavitations as dental trauma, bacterial infections or toxicity. Dental trauma includes extractions, anesthetic injections, root canals, gum and bone surgery, tooth grinding and clenching, excess heat from drilling and electrical trauma from placement of dissimilar metals. Bacterial infections may arise from abscessed teeth, periodontal (gum) infections, bacteria harbored in the dentinal tubules of teeth with root canals, infected wisdom teeth and cysts. The causes of toxicity include dental filling materials such as mercury amalgam, root canal filling materials, anesthetic by-products and the vasoconstrictors used in the anesthetic, bacterial toxins, chemical toxins and drugs such as cortisone.

Diagnosis of cavitations is made from patient history and symptoms, panoramic x-rays, bone scans, blood analysis, Electro Dermal screening, and an instrument that measures bone density called the CAVITAT.

The recommended treatment for cavitation lesions of the jaws is primarily surgery and some nutritional and homeopathic remedies.

The prognosis or outcome of cavitational surgery is bleak. Even the best and most experienced surgeons expect a 40 to 50% failure rate. Repeated cavitational surgery is very common, with some patients reporting as many as 36 or more cavitational surgeries. In some cases, surgery has been repeated as many as four times on the same site without satisfactory results. There are no long-range studies that show cavitational surgery extends life. Moreover, diagnosis of cavitations is often flawed due to the use of a single diagnostic approach or misinterpretation of the technique.

Let's first examine the panoramic x-ray. All x-ray interpretation depends upon the density of the bone; the harder or denser the bone, the more white or radio opaqueness shows on the x-ray. Dense bone has more bone cells. Conversely, less dense bone has fewer large bone cells and looks more black or radiolucent on x-rays. It is a known fact that bone that is not loaded or stressed, regardless of its location in the body, becomes more radiolucent. That is, it loses bone cells, and the remaining bone cells have thinner walls with less blood supply. An example everyone can understand is the hipbone. Less active or non-ambulatory people lose bone density in the hips, thus leading to osteoporosis with more risk of fracture. Extraction of the teeth results in the same loss of bone density due to the unloading of the jawbone. Therefore, it stands to reason that tooth extraction causes a decrease in bone density and poorer circulation that will show up on a panoramic x-ray as a "black hole." But this type of "black hole" is an incomplete and inconclusive way of diagnosing cavitations.

The CAVITAT instrument works on the same principle of bone density as the x-ray. The instrument registers bone that is less dense. There are many non-pathologic reasons for bone to be less dense. Previous extraction sites, sinuses, medullar nutrient channels, nerve channels and hereditary factors might give a person the light or slight bone features. One could arrive at the same conclusion as with x-rays. Therefore, the CAVITAT, by itself, is an incomplete and inconclusive way of diagnosing cavitations.

Electro-Dermal screening devices are computerized models of the early Electro Acupuncture by Voll, commonly called EAV machines. Many companies now make computerized EAV units. There is no question that these units can accurately measure energy coming off the body, in much the same way that the heart function can be measured with an EKG or the brain energy measured with an EEG. The big question has to do with the possibility of the operator influencing results and the accuracy of the original designation of particular indications (or readings) of the device as "pathological."

Chinese acupuncturists have known for centuries that the teeth control the meridians, but what may or may not be wrong with the teeth is still open to debate. Electro-Dermal screening devices that measure altered energy due to dental conditions may have been incorrectly programmed. An indication that the energy interruption is due to mercury or other toxic filling materials, root canals and/ or infection in the bone, may just be an energy interruption based upon the proprioceptive feedback mechanism to the brain from the muscles in the head, neck and jaw area.

There is mounting evidence that "intent" plays a role in diagnosis and "fear" plays an even bigger role in getting well. A word of CAUTION! <u>Do not plunge</u> into tooth extraction and cavitation surgery based upon limited diagnostic methods or one opinion from a dentist or other medical practitioner. While extracted teeth and lost bone can be replaced, the surgery is lengthy and costly compared to saving your own teeth and bone.

Our society has been led to believe that illness and ill health result from a "cause and effect" relationship. We are always looking for a new microbe or virus, a genetic flaw, a toxin or something to blame for our illnesses. We could benefit from studying those who do not get sick from mercury fillings, root canals or cavitations. Millions of people around the world live longer and with less medical expenses than Americans. How do they do it? Obviously, there is no singular factor, but let us investigate proprioception to the brain as a possibility.

In a previous chapter I have presented an in depth discussion of the role of faulty proprioception in many ailments. It also is a fact that proprioception to the brain is not well understood by dentists who recommend the removal of teeth and perform cavitational surgery. Several studies have shown that loss of the posterior (back) teeth, or even loss of the height of the back teeth, leads to faulty proprioception to the brain and an imbalance of the Sympathetic and Parasympathetic Nervous Systems.

Since cavitation surgery is an invasive process with many post-operative symptoms and a poor prognosis, let's examine a non-invasive and inexpensive way to treat suspected cavitations. This non-invasive cavitation treatment protocol involves placing a Proprioceptive Guide over the lower back teeth, and the use of low level laser therapy. There are several advantages to this approach:

- It can be done by either a professional or by a layperson in their own home.
- It is non-invasive.
- It is inexpensive compared to the cost of surgery.
- Results can be monitored by improvement in the same

tests used to diagnose cavitations or by improvement of symptoms attributed to the root canals or cavitations.
- Results are quickly seen in 2 to 4 weeks.
- If results are not seen, cavitation surgery can still be used as a last resort.

If your dentist has an open mind and is willing to listen to you, request that he/she take an impression of your upper and lower jaws and make splints that fit over your lower back teeth. Upper splints don't work for changing faulty dental proprioception. The splint must be on the lower arch.

If you have already lost your lower back teeth, make combination tooth/gum supported splints. NOTE! It is very important that these splints do not have a front that allows the front teeth to hit when closing, nor should the splints have any indentations or bite for the upper teeth to fit into. (Night guards won't work!) These splints should be totally flat and slightly higher on your sore side.

If you cannot find a dentist, have confidence in your own ability, and have not lost your lower back teeth (excluding wisdom teeth or 3rd molars), you can order a Miracle Bite Tab kit from GO-JO Enterprises (605-342-5669) and make your own Proprioceptive Guide as mentioned in the previous chapter. Remember, MBTs are not considered a permanent solution and should be used as a "band aid" along with the laser to get you out of trouble.

My protocol for treating cavitations is similar to treating root canals. The Proprioceptive guides MUST be adjusted according to instructions and worn and at all times, except when eating. Use the 660 Enhancer in the mouth directly over the gum of the suspected cavitation for six breathes. Follow this with the 808 Enhancer over the "hole in the bone" through the skin (outside the mouth) for six breathes. This procedure should be repeated three times per week for two weeks, and then twice a week until symptoms are gone or tests show improvement.

This simple non-invasive Proprioceptive/laser technique has saved many people from what can be painful and destructive cavitation surgery, and it has restored their quality of life. Remember, in the animal world all creatures die when they loose their teeth. SAVE YOURS!

8.7   Low Level Laser Therapy for Other Dental Conditions

Dental treatments that benefit from the combined use of low level laser therapy are wound healing, aphthous stomatitis or mouth ulcers, pulpotomy, mucositis, neuronal regeneration and post-herpetic neuralgia. Other areas showing some promise in the use of low level laser therapy include synovitis, acute abscesses, periapical granulomas, chronic oral-facial pain, and post-extraction pain.[36]

Application of low level laser therapy has enhanced bone growth around dental implants in animal models. The findings are encouraging. However, low level laser therapy's effect on human hard tissues requires more studies.

With more studies being performed every year in the dental field, it is reasonable to assume that more conditions will respond to the treatment of low level laser therapy. My book, *Low Level Laser User's Manual,* includes several protocols for treating dental conditions such as canker sores, cracked lips, dentine sensitivity, gingival retraction, bleeding gums, inflammation, periodontal disease, and sensitive teeth. I have also provided a protocol for anesthesia. I hope that the many possible applications of low level laser therapy will help you care better for your teeth.

Chapter 9   Emergence of an Affordable Health Care System

9.1   Medical Science Meets the Practical Need for Home Health Care Maintenance

Science exists at two levels, the speculative, abstract level and the practical, concrete level. We humans, with our limitless curiosity, feel a need for knowledge, and seek to understand our world. This psychological need, a product of our intelligence, is natural. The scientific method is one of the great human achievements which emerged from the intermingling of ideas from many civilizations. This method dictates the need for rigorous experimentation and evidence to substantiate the investigations that our intuitive hypotheses lead us to propose. The history of science makes us aware that our present views are merely theories, temporary conclusions that will no doubt be replaced by future research and experimental evidence. Remember, the consensus of medicine has never been right. The field of medicine in particular is always changing, example, in my youth it was common practice to have your tonsils removed. Today tonsillectomies are a last resort. These theories, which evolve into unconsciously accepted all-pervading cultural systems, are sometimes referred to as "paradigms." Many of us have learned to approach all scientific claims with a healthy skepticism. Although often competitive and not always ethical and above-board, science is a grand project where we work together and share in the process, constantly fine-tuning the results. The need to criticize and challenge one another's conclusions, referred to as the peer review process, helps to hone our theories and knowledge.

The second, more concrete level is important because it is the way that science filters down to the layperson and in fact hits us in the pocket book, whether on the personal or the national level. The abstract principles of science are applied in such areas as military programs, space projects and of most concern to us here, in medicine. We often hear about the "crisis in medicine." This means that the above-mentioned creative and imaginative level of medical science has led to the application of "brilliant" speculative ideas in the production of prohibitively expensive devices both for diagnostics and treatment. Moreover, the requirement for peer reviewed controls and "protections" has raised the price of medicine and synthetic drugs because of the exorbitant cost of research and testing and the litigation associated with medical mistakes and failures. Therefore, both the medical technology and the medicines have become so expensive that they strain the budgets of individuals and families, as well as hospitals and, of course, the public health care system. I used to think that medical/dental insurance companies "ruled." They have the biggest buildings, employee millions of people, process billions of claims and of course make billions of dollars --- so why are they going bankrupt?

The Reliance Insurance Company, one of the largest insurers of dentists, and the carrier for the American Dental Association (ADA) went bankrupt in the late 1990s. Why is that? In the 60s and 70s dentists seldom got sued even though they were working in one of the most delicate areas of the body, the mouth. During this era, if a dentist got sued, regardless of the validity of the charge, it was "settled out of court." Reliance Insurance Company routinely paid $20,000 to $50,000 to settle out of court, even though the case had little or no validity, just to keep the dentist's name out of the newspapers. Attorneys quickly picked up on this insurance company error. Even if they knew a complaining dental patient had little or no chance to win a law suit, they took the case and filed it–because they knew it was an automatic payoff by Reliance. Attorneys received their commission fee and the dentists' malpractice insurance went up, but not rapidly enough to save Reliance from bankruptcy.

However, despite this tension between creativity and practicality, the historical process that maintains balance is in action right now. Medical science has gone to such an extreme of specialization, that there is a sense that the concern for the patient as a whole person is lost. Obviously, change is needed and, fortunately, a natural swing is occurring in the paradigm of medical science. These circumstances require that we go back to the drawing board and rethink our premises and procedures, consider more carefully the importance of body terrain and mind-body functions in order to keep ourselves healthy. Many creative minds have done just that and have come up with new solutions and systems; the current understanding of the use of light and electromagnetic energy has made many things possible. We all know about the new imaging technology that allows doctors to see inside the human body without exploratory surgery. Ironically, some of this top notch, high level theoretical principles of light and energy have been applied in devices that can be used at home by the layperson. These tools do not need to filter through expensive medical systems requiring multiple doctor's visits, hospital stays and pharmaceuticals. Those working in the field of "holistic and alternative" medicine have reached into the bag and come up with simpler (not to be confused with "simplistic") systems based on different perceptions of the cause of illness and bodily dysfunction. They emphasize prevention, and enhancement of the body's natural tendencies.

9.2   Miracle Bite Tabs

You have read about dental distress syndrome and faulty proprioception to the brain caused by the long-term effects of modern dietary changes on the structure of our mouths and teeth. Now a simple device— Miracle Bite Tabs™ (MBTs)—helps alleviate that chronic condition. Miracle Bite Tabs™ aid patients coping with persistent, debilitating health problems, often when no other medical treatment has been effective. They combine existing polymer technology with ancient Chinese medical theory. You have also read about the crucial importance of a well-functioning digestive system. I have created another product, *Belly Gelly®*, to help digestion and promote natural detoxification.

9.3   Another Creation from Dr. Larry Lytle, "Belly Gelly®"

Eating is a complex process. When the gastrointestinal system functions properly, our body absorbs the nutrients it needs to maintain our health. But even if we eat fresh, wholesome food, our bodies cannot process it completely if our intestines are full of accumulated toxins and compacted wastes. The intestinal flora and surface area must remain unimpeded. A product called *Belly Gelly®* provides an inexpensive and simple way for patients to detoxify and cleanse their digestive tracts so that they get the most out of the foods they eat. Regular ingestion of *Belly Gelly®* helps propel one another step up the ladder towards total health.

The history of *Belly Gelly®* goes back to a pioneering California colon hydrotherapist named Edward Irons, Ph.D., and author of a book called *Death Begins in the Colon,* which is often quoted by holistic practitioners.[94] Dr. Irons also developed a bowel-stimulating product generically identified as Bentonite suspension.

Bentonite is actually an ancient volcanic ash, which is comprised of a mixture of clay and minerals. It absorbs most things it comes into contact with but still remains inert. Some bentonite products possess trace minerals and others merely consist of clay. Dr. Irons used his bentonite suspension himself and recommended it to thousands of people when he lectured before audiences attending two California based, holistic-oriented organizations, the National Health Federation and the Cancer Control Society. This colon hydrotherapist, who lived to be nearly 100 years old and who fathered a son when he was in his early 90s, credited his excellent health, potency, and longevity to his careful daily colon care with the bentonite suspension product that he manufactured.

I used Dr. Irons' bentonite for colon health myself and recommended it to many people who benefited from it during my clinical nutritional practice. I reasoned that if bentonite absorbed nine times its weight, then a heavier substance would absorb even more. This reasoning led to the development of a completely new product, *Belly Gelly®*. Although *Belly Gelly®* was effective, distribution was difficult in the beginning because users did not care for the taste. But after many experiments, which included the addition of natural ingredients and the use of a special processing technique, the consistency and taste of *Belly Gelly®* improved to the extent that it was readily accepted by the public.

Today, health professionals emphasize the importance of a healthy immune system. Chronic immune system disorders such as lupus erythematosus, fibromyalgia, Crohn's disease, multiple sclerosis, Lou Gehrig's disease, AIDS, and many others, are destroying millions of lives. But few people understand the immune system and why it is often under active.

In part, the immune system is composed of various white blood cells. One might wonder where these defensive blood cells are produced. Most people believe white blood cells are created in certain organs such as the thymus gland and spleen, and in the long bones. But that is only partially true. About seventy-five percent of the life-saving white blood cells that constitute the immune system are actually manufactured in the Gut-Associated Lymphoid Tissue (GALT) which lines the intestines. If spread out, the entire absorptive area of an individual's gut would be larger than a football field. The quality of this tissue is of major importance because this is where a person's immune system begins its function; unfortunately, most people are highly toxic right there—in the gut. One autopsy study discovered that from 7.5 to 75 pounds of encrusted fecal material line the colon of most people. The famous western film actor, John Wayne, was rumored to possess 70 pounds of encrusted toxins in his gut at the time of his death from cancer.

It just makes good sense, therefore, to follow the advice of Dr. Irons—keep a healthy colon. There are hundreds of detoxification programs, techniques, and products on the market today. Every conscientious nutritional counselor probably recommends intestinal detoxification to clients. However, some detoxifiers work well and others do not. To enhance the immune system and increase production of our body's life saving immune cells, use *Belly Gelly®* in conjunction with the Q1000 laser; apply mode 3 for one cycle, or three minutes each, on the ascending, transverse and descending colon.

Often, problems arise in the gut from taking excess expansion fiber supplements sold in health food stores. Peristaltic action, involving a systematic constriction and release of the smooth muscles, is a mechanism that moves the waste through the gut. This muscle action is stimulated by the vagus nerve, which receives its signal from the proprioception to the brain from the dental muscles. Failure of this system to function properly can lead to constriction of the smooth muscles and a blocked bowel. Since dietary fibers expand and absorb water, the blockage can become a serious problem. Therefore, care should be taken not to ingest dietary fibers when the bowel has lost its peristaltic action. Peristaltic action, a Parasympathetic Nervous System function, can be normalized by raising the height of the back teeth with some type of dental splint such as Miracle Bite Tabs.

Belly Gelly® does not expand; it remains slick as it passes through the gut. Proven to be one of the easiest, safest, and most effective bowel detoxifiers that one can ingest, *Belly Gelly®* can be useful as an antidote for diarrhea, constipation, nausea, and for prevention of bowel upset.

Belly Gelly® does absorb everything in its path. Therefore, the *Belly Gelly®* program definitely requires a re-inoculation of healthy and friendly bacteria or probiotics into the gastrointestinal track. I recommend one probiotic in particular, that does not break down in the stomach acid. Geneflora®, also called Bio-Plus, is a favorite of mine because it does not become activated until it reaches the intestines where the internal pH is 8.

In vitro laboratory research has demonstrated that low level laser therapy can enhance the growth and production of the cellular components of the immune system. While this assertion may be difficult to prove *in vivo* (in the body), it makes good sense to laser the gut with mode 3 (multi-organ mode) of the Q1000 laser. The proven safety of this laser has already been discussed.

9.4 Three Economical Products for Self Care

Miracle Bite Tabs™ and *Belly Gelly®* have been integrated with low level laser therapy into a system for effective home health care maintenance. The *Q Series Low Level Lasers* have been proven to stimulate the body's inherent ability to heal itself. The results are observable and contrast dramatically with typical healing scenarios. Unlike most medical procedures, low level laser therapy does not treat symptoms directly; instead, through the process called biostimulation, it naturally energizes the body's cells.

These three products, which can be self-administered by the patient, are not substitutes for proper medical supervision. Pain is an indication that the body is not functioning correctly. Medical advice should always be obtained in a timely manner. An illness is usually more easily cured during the early stages of the disease process. Therefore, early detection is much better than ignoring the pain until it is excruciating. The devices are considered to be "complementary medicine," and may be used in conjunction with treatment prescribed by a doctor.

On the other hand, if the doctor does not detect any life-threatening illness, these devices may be just the right ingredients needed to boost the body's inherent capacity to heal itself. An important statement often repeated by practitioners of CAM claims that: "Helping the body to heal itself in every way we can, is what makes total health attainable."

Although, at first glance the purchase of a hand-held laser for $3795 seems expensive, over time you will find that the money is well spent. The following three testimonials support the use of low level laser therapy as an economical solution in the current health care crisis:

I have used a low level laser - Q1000 - since June 2002. This is the most profound therapy I have ever experienced. In every single condition for which I use this miraculous device - cardiac conditions, gastrointestinal function, gynecological problems, immune system modulation, microcirculation, muscle regeneration, osteoarthritis, tendonitis, and pain associated with any cause - healing occurs and the condition dramatically improves beginning on the next day after the initial laser therapy treatment. *Dr. Nataliya Rakowsky, M.D. of St. Petersburg, FL*

My New Q1000 Paid For Itself In One Weekend! Last month, I purchased a Q1000 laser from Doug Phillips after my wife received a demonstration of this amazing product which, within 30 seconds, eliminated a severe headache she was experiencing. Three days ago, on Friday, May 21, 2004, I was kicked extremely hard by a 1300-pound horse. The hoof struck and smashed my left hand, nearly imbedded a metal bucket I was carrying into my left leg, and drove my entire body to the ground with incredible force. When I got up, my first thought was that with an injury of this severity I'd have to close my practice for a couple of weeks. I have a busy practice and the thought of losing one or even two weeks of income to an injury … well, it's just something that I'd rather not think about. I immediately iced my hand for about 15 minutes and then applied the Q1000 to both my hand and leg. (Thankfully, I had listened to Doug when he had told me to take my laser with me everywhere I go!) I used the Q1000 again on both areas later that evening. The next morning I awoke and discovered that I had been laying on the injured hand a little during the night, and was surprised that the pain was 95% gone! Today, I am at work helping my patients towards better health, and have only a slight swelling of the middle finger. If you'd asked me Friday evening if I would be at work on Monday morning, I would have told you: 'no way!' I never imagined that I'd be able to recover so quickly … especially from an injury of the hand that I use in my practice on a daily basis. The Q1000, in just this one situation, has literally saved me thousands of dollars! I have also recently used the Q1000 on my wife, a patient and my assistant – all of whom had received very bad sunburns. All three received tremendous pain reduction benefits from the Q1000...My wife's pain was nearly gone in only 24 hours! *Dr. Terry Shroyer, Junction City, KS*

We use the Q1000 in the early stages of colds and fever blisters, and enjoy a rapid reversal of symptoms. Headaches don't' linger and wounds heal quickly. Our son (16 years of age) applied regular laser treatments on a gash on his forehead (a wrestling injury) with scar-free results! We are armed with a powerful healing tool in the event of future health challenges. We highly recommend the laser and enhancers! *Kitty Nordstrom, Costa Mesa, CA.*

Chapter 10   Conclusion

10.1   The Therapeutic Benefits of Laser Light Result From Biostimulation on the Cellular Level

The Q1000 laser, the result of decades of research, provides a method for adding energy to the body and stimulating cellular activity. This low level laser light tuned to specific frequencies (wavelengths), can stimulate metabolic processes in the human body at the cellular level. More than 100 positive double-blind studies, conducted throughout the world, attest to the fact that this laser light stimulates healing.

This low level laser emits light wavelengths that bring about observable health benefits. First-hand patient experience has shown that the therapeutic benefits result from:

- increased blood circulation to injuries
- reduction of the incidence of infection and infectious organisms.
- diffusion of inflammation.
- relief of pain.
- stimulation of cellular activity and growth in soft tissues such as skin cells and hard tissues such as bone.

The laser probes can also stimulate acupuncture points in a simpler and faster way than traditional acupuncture needles utilized in Traditional Chinese Medicine. In fact, needles are not needed. The laser light and the probes can also be used in dentistry. Most importantly, these lasers are simple enough for patients to safely treat themselves and thus avoid unpleasant and costly medical visits.

Good nutrition and regular detoxification, in conjunction with the low level laser therapy, can contribute to effective and affordable home health care. Given the looming crisis in our public health care system, this is important. Prevention and early intervention can make the difference in saving your own life and the lives of your loved ones

10.2   New Research Findings: More Uses of Low Level Laser Therapy on the Horizon

The rapid progress of low level laser science and technology suggests the proliferation of laser use in the near future. For example, biomedical researchers in the country of Colombia irradiated adipose tissue (fat cells) which had been removed from patients receiving lipectomies (fat reduction surgery). They observed that, "After 4 minutes of laser exposure, 80 percent of the fat was released from the adipose cells; at 6 minutes of laser exposure, 99 percent of the fat was released from the adipocyte (fat cells). The released fat was collected in the interstitial space (outside the cells)....The low-level laser energy affected the adipose cell by causing a transitory pore (temporary opening) in the cell membrane to open, which permitted the fat content to go from inside to outside the cell."[95]   In other words, low level laser therapy can break up fat, which is a highly attractive characteristic of treatment and the "holy grail" for millions of overweight people.

Doctors in Turkey found that patients suffering from fibromyalgia responded well to low level laser therapy. *Fibromyalgia syndrome* (FMS) is a muscular and skeletal pain and fatigue disorder for which the cause is still unknown. FMS patients experience terrible discomfort in their muscles, ligaments and tendons, the body's fibrous tissues. FMS patients say that they "ache all over," and their muscles feel like they have been overstretched or overworked. They also feel devoid of energy, similar to those with chronic fatigue syndrome.

After treating FMS patients with low level laser therapy, the Turkish physicians observed that, "Significant improvements were indicated in all clinical parameters (disease symptoms) in the laser group".[96] These improvements included: reduced pain; reduced number of tenderness areas; less joint stiffness in the morning; more restful sleep; less muscle spasm; and less fatigue.

A recent dissertation by A. S. G. Segundo of the University of São Paulo, Brazil, indicates that when low level laser is used in conjunction with the photosensitizer, azulene, the bacterial counts in root canals can be significantly reduced. Photodynamic therapy is a procedure in which a dye is excited with an appropriate wavelength of light. The excited photosensitized dye reacts with oxygen to form the highly reactive compound, singlet oxygen, which kills bacteria and tumor cells. This study evaluated the bacterial reduction in root canal contaminated with *Enterococcus Faecalis*. The azulene dye was irradiated with a 685nm diode laser with a power output of 10mW for 3 minutes.[97]

Near infrared light therapy, also known as low level laser therapy, is drawing a lot of attention from research clinicians around the world. For a number of years, various research centers in Japan, Britain, and the United States have been conducting clinical trials to measure the efficacy of the application of red and near infrared light over injuries and lesions. The results show that they contribute to healing and provide relief for both acute and chronic pain.

Many of these trials have proven to be very successful and clearly verify that light can have a positive effect on damaged cells. The results are so outstanding that the US Defense Advanced Research Projects Agency is funding research into the method. The agency hopes to use light to treat personnel whose eyes have been damaged by high powered lasers during combat.

Scientists have put light emitting diodes (LEDs) to the test on eye injuries. In a study that will appear in *The Proceedings of the*

National Academy of Sciences, Dr. Whelan blinded rats by giving them high doses of methanol (wood alcohol). Within hours, the rats' energy-hungry retinal cells and optic nerves began to die. Within one to two days the animals went completely blind. If the rats were treated with LED light, at a wavelength of 670 nanometers for 105 seconds, at five, twenty-five and fifty hours after being dosed with the methanol, they recovered 95 percent of their sight. Remarkably, the retinas of these rats looked indistinguishable from those of normal rats. Whelan states: "There was some tissue regeneration, and neurons, axons and dendrites may also be reconnecting."[98]

The president of the North American Association for Laser Therapy, Dr. Juanita Anders, has been engaged in laser therapy research for many years. Dr. Anders and her team, at the Uniformed Services University of the Health Science in Bethesda, Maryland, have documented a number of clinical trials which verify the positive effects of laser therapy on neuron regeneration following injury.[99] Recent studies have shown that laser therapy has the ability to inhibit inflammatory cell invasion and activation in the spinal cord. Dr. Anders' current research is focused on spinal cord problems; wound healing in diabetes, and the treatment of axotomized facial motor neurons. -

The FDA approved a laser called N-Lite for acne treatment in June, 2003. This type of laser has been used in other countries for several years to treat wrinkles, scars, and other lesions. It has now been approved as an effective device for the most common skin problem of all—acne. This skin problem affects about 85% of Americans at some point in their lives. Dr. Tony Chu and his colleagues in Britain conducted a study on acne in 31 patients. Their report in *The Lancet* states that after just one 15-minute treatment, the overall severity of acne was reduced by one half.[100] The control group (10 patients), which received a sham laser treatment, got little improvement. The patients were observed for three months after treatment and the acne was still under control with no adverse effects. This is a very encouraging outcome. The laser not only reduces the inflammation associated with acne, it also helps in the production of collagen. The latter effect makes the laser useful for removing wrinkles and scars—its primary applications in this country since 2000, when it was first approved for use. N-Lite has also been used to treat psoriasis, rosacea, eczema, and warts, and is currently employed in some 200 dermatology practices in the U.S.

For several years, the so-called "blue lights" have been used to kill bacteria. However, because they don't penetrate the skin as deeply as N-Lite, the results for acne are short-lived. More powerful lasers have also been used on acne, but only on the back where the skin is tougher because they can cause irritation. Since N-Lite has a wavelength that doesn't damage skin, it can be used on facial skin and is relatively painless. There is also a prescription medication for severe acne, called Accutane, but not only is it expensive, it can also cause serious side effects. Laser seems to be a good alternative. A single N-Lite treatment costs about $600 in the U.S. For a few treatments of N-Lite at a dermatological clinic, you would be able buy your own Q1000 laser device and safely treat your acne at home. You would then have the laser to use not only for acne but whenever needed for other conditions proven to be responsive to the treatment of low level laser therapy.

The following study illustrates the benefit of owning one's own laser. Dr. Colin Carati of Flinders University in Australia conducted a study using low level laser on women suffering from lymphedema. This is a chronic and progressive condition with few effective treatment options that is often caused by mastectomy in breast cancer patients. In the trial, 61 patients were assigned to one or two cycles of either genuine laser therapy or a sham therapy with a disabled laser. The treatment did not have an immediate effect on symptoms but two to three months later, those who had two cycles of actual treatment, found that their swelling was reduced by 31 percent. The skin on their upper arms was also softer. (Hardening of the skin is a side effect of lymphedema.)[101] However, the therapy did not improve the range of movement in the arm, the quality of life, and the ability to perform daily activities. These outcomes point to the possibility that more frequent usage and longer duration of treatments, might be even more beneficial for lymphedema than was observed in the above-mentioned study.

10.3   Low Level Laser Therapy: An Important Complement to Sports Medicine

It was recently reported in the Washington Post that players of the 2004 National Football League champion team, the New England Patriots, used laser for pain relief and wound healing. It has also been reported that Lance Armstrong, the six-time champion of the *"Tour de France,"* uses low level laser. Many of our Olympic teams are also adopting laser therapy for their players. Our laser users experience the same results. Here are two testimonials:

> My daughter, Whitney, had a severe wrist inflammation in 2003. She had pain in her wrist from overusing it on the tennis court. She could not play tennis without pain. Whitney received one demonstration with the Q1000 and 808 Enhancer probe and her wrist felt better almost immediately. Even after playing tennis at a 100% level, the pain never returned. Whitney ended up winning the 2003 South Dakota high school tennis singles title. I would like to thank Doug Phillips for demonstrating low level laser therapy to Whitney and me. *Daryl Paluch, USPTA Tennis Professional, Rapid City, SD*

> My son is a member of his high school track team. During one important meet, we noticed that a lot of the athletes were in some form of pain. We decided to use the Q100 on our son (and) some of his teammates. The laser made all the difference in how they finished their events. *Frances Brown, Seattle Washington*

The ability of low level laser therapy to relieve pain, perform acupuncture, dissolve fat, regenerate bone tissue, rejuvenate the skin, and accomplish many other things, truly makes it a wonderful treatment, on its own, or in conjunction with other therapies.

10.4 Low Level Laser Therapy for Promotion of "Total Health"—Not Just Disease Control

"Total health" is not the absence of disease, but rather complete mental, physical, structural, dental and spiritual well-being. This is a positive state of homeostasis where the body functions optimally and is able to prevent disease itself. When both body and mind are healthy, the individual feels energetic and psychologically motivated to accomplish their goals. They exhibit physical strength and dexterity sufficient to perform the physical challenges of their jobs and their sports activities, as well as deal with the stress of everyday life. They excel at whatever they do.

We have talked extensively about the Autonomic Nervous System, proprioception to the brain, and the beneficial effect of low level laser therapy in normalizing the Autonomic Nervous System and correcting faulty proprioception. Forthcoming clinical evidence indeed corroborates these benefits.

Dr. Herbert Yolin's presentation at the 2004 Annual Conference of the Academy of Laser Dentistry demonstrated the use of heart rate variability (HRV) testing to gauge the effect of low level laser therapy on stress and the Autonomic Nervous System. HRV measures the functionality of the Autonomic Nervous System and is a favorite functional assessment tool for some of the most progressive integrative medical doctors and health practitioners. HRV is a good non-invasive indicator of one's state of health. Little variation in heart rate, when in a stressed state compared to a normal state, indicates the adaptability of the Autonomic Nervous System. Good heart rate variability means you have enough reserve in your Parasympathetic Nervous System to encounter any stressful situation. After treating 100 patients with low level laser therapy, Dr. Yolin found that their stress responses improved from 24 to 72%. This indicates a substantial improvement of the Autonomic Nervous System balance, which can lead to improvement of overall health.

If you have no symptoms and have not been diagnosed with any illness, low level laser therapy can still contribute to your health by increasing your Parasympathetic Nervous System reserve and helping you achieve balance in your Autonomic Nervous System. Perhaps it can even deter the process of aging.

Another line of evidence in supporting the use of low level laser therapy comes from the imaging data of thermography, which can detect any inflammation in the body, including that in the TMD/TMJ area. In a few case reports we have seen reduction in inflammation immediately following the application of low level laser. Many disease states involve chronic inflammation. If inflammation is immediately reduced following treatment with low level laser therapy imagine the benefits of receiving treatments more often. It is very clear that continuous application of low level laser therapy at home will help contain disease (Figure 33)

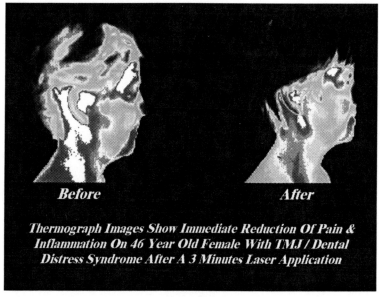

Before   **After**

Thermograph Images Show Immediate Reduction Of Pain & Inflammation On 46 Year Old Female With TMJ / Dental Distress Syndrome After A 3 Minutes Laser Application

Fig. 33

The many benefits of low level laser therapy will be widely recognized within a few more years, much because of the gradual accumulation of solid clinical studies proving its contribution to the care of disease and its significance in the attainment of "total health" and wellness. Many people remain skeptical of low level laser therapy and are locked into the mind-set of medicine as "disease care." (This means that people who are symptom free don't see the need for preventive action and maintenance of "total health" and wellness.) Fortunately, a gradual paradigm shift, that encompasses a concept of "total health" and the body's innate capacity to heal itself in the right conditions, is occurring throughout North America and in the wider world. Low level laser therapy is an important component in the newly developing technology and philosophy underlying holistic medicine.

Resources

For more information about low level laser therapy, location and dates of my **Day of Learning Seminars** and **Healing Light Seminars**, please see the website www.laserinformation.com or send your e-mail query to lytle@pobox.com or telephone to Low Level Laser Consultant, Dr. Larry Lytle - (605) 342-5669, or fax (605) 342-5739.

For more information or to purchase the Q Laser Series, call 1-800-666-3774 for the name of a Distributor near you.

To purchase products discussed in *Healing Light*, visit www.laserinformation.com;

e-mail lytle@pobox.com ; call 1-866-375-9853

To purchase Dr Lytle's books, tapes, and videos, visit www.drlytle.com,

e-mail gojo@pobox.com, or call 1-605-342-5669 or 866-375-9853

     Healing Light – Dr Larry Lytle

     Low Level Laser User's Manual - Dr Larry Lytle

     Understanding Low Level Laser Therapy – Dr *Larry Lytle*

     Low Level Laser Physics – Dr Larry Lytle (30 minute audio tape presented at the 2003 Cancer Control Society annual meeting).

Healing Light Workshop Video Library – Dr Larry Lytle

For Additional Reading – (To Order Call 1-800-666-3774 or e-mail gojo@pobox.com)

High Tech Pain Management For Pets – Dr Tamara Shearer

Miracle Bite Tab Instructional Video –Dr Herb Yolin

Mind Map – Buddy Frumker

The Dental Physician – Aelred C.Fonder

Laser Therapy – Jan Tuner and Lars Hode

Lasers in Medicine and Dentistry and Medicine – Zlatko Simunovic, plus contributions from 25 of the World's leading laser authors

References

[1]Fonder, A. C., *The Dental Physician*. (Rock Falls, IL: Medical Dental Arts, 1985).

[2] McTaggart, L., *The Field: The Quest for the Secret Force of the Universe*. (New York: Quill, 2003) pp. 39-43; 48-49.

[3] McTaggart 48-49.

[4] *Holography: Commemorating the 90th. Anniversary of Dennis Gabor:* 2-5 June 1990, (Tatabanya, Hungary, Institute Series, vol. 08) by Institute on Holography, *et. al.*

[5] Oschman, J., *Energy Medicine in Therapeutics and Human Performance*. (Philadelphia, PA: Butterworth Heinemann, 2003), pp. 172.

[6] Oschman, pp. 200.

[7] www.psy.cmu.edu/~davia/mbc/8start.html

[8] Oschman, pp. 250.

[9] Laser Classification--www.ANSI Z136.1-1993

[10] Laser Classification--www.uos.harvard.edu/ehs/radsafety/las_war.shtml

[11] Laser Classification, Code of Federal Regulations, Title 21, Vol.8, Sec. 1040.10.

[12] Durnov, L.A.; Gusev, L.I.; Balakirev, S.A., *et al.* "Low-intensity Lasers in Pediatric Oncology." *Vestn Ross Akad Med* (6): 24-27, Nauk. 2000.

[13] Balakirev, S.A.; Gusev, L.I.; Kazanova, M.B., *et al.* "Nizkaointensivnaia lazernaia terapiia v detskoi onkologii." *Voprosy onkologii* 46(4): 459-461, 2000.

[14] JAMA (*The Journal of the American Medical Association*, July 26, 2000, 284, 483-485)

[15] Brugnera, A.; da Silva, OP.S.; de Vieira, A.L.; do Nascliento, S.C.; Pinheiro, A.L.; Rolim, A.B., "Does LLLT Stimulate Laryngeal Carcinoma Cells? An *in vitro* study." *Brazilian Journal of Dentistry* 13(2): 109-112, 2002.

[16] Chernova, G.V. and Vorsobina, N.V. "Effect of Low Intensity Pulsed Laser Radiation of Basic Parameters of Aging in *Drosophila Melanogaster.*" *Radiats Biolog. Radioecol.* 42(3): 331-336, May/June 2002.

[17] Lytle, Larry, D.D.S., Ph.D., *Understanding Low Level Laser Therapy.*

[18] Branco, K.F.; Hahn, K.A.K.; Lieberman; B.E.; Naeser, M.A. "Carpal tunnel syndrome pain treated with low-level laser and microamperes transcutaneous electric nerve stimulation: A controlled study." *Archives of Physical Medicine and Rehabilitation* 83(7):978-988, July 2002.

[19] Mester, A.; Mester, E.; Mester, A. "Open Wound Healing—Bed Sores, Ulcus Cruris, Burns—with Systemic Effects of LLLT." In *Lasers in Medicine and Dentistry: Basic Science and Up-to-Date Clinical Application of Low Energy-Level Laser Therapy—LLLT* edited by Zlatko Simunovic, M.D., F.M.H. (Zagreb, Croatia: Agencija za komercijalnu djelatnost, March 2000), pp. 227-244.

[20] Veen, P. and Lievens, P. "Low Level Laser Therapy (LLLT): the Influence on the Proliferation of Fibroblasts and the Influence on the Regeneration Process of Lymphatic, Muscular and Cartilage Tissue." In *Lasers in Medicine and Dentistry: Basic Science and Up-to-Date Clinical Application of Low Energy-Level Laser Therapy—LLLT* edited by Zlatko Simunovic, M.D., F.M.H. (Zagreb, Croatia: Agencija za komercijalnu djelatnost, March 2000), pp. 187-215.

[21] Simunovic, Z. "The Healing of Post-operative Wounds with LLLT." In *Lasers in Medicine and Dentistry: Basic Science and Up-to-Date Clinical Application of Low Energy-Level Laser Therapy—LLLT* edited by Zlatko Simunovic, M.D., F.M.H. (Zagreb, Croatia: Agencija za komercijalnu djelatnost, March 2000), pp. 245-250.

[22] Tsourouktsoglou, A.T. "The Use of LLLT in Wound Healing in Greece." In *Lasers in Medicine and Dentistry: Basic Science and Up-to-Date Clinical Application of Low Energy-Level Laser Therapy—LLLT* edited by Zlatko Simunovic, M.D., F.M.H. (Zagreb, Croatia: Agencija za komercijalnu djelatnost, March 2000), pp. 251-253.

[23] Bensadoun, R. "Low Energy He-Nelaser in the Prevention of Radiation-induced Mucositis." In *Lasers in Medicine and Dentistry: Basic Science and Up-to-Date Clinical Application of Low Energy-Level Laser Therapy—LLLT* edited by Zlatko Simunovic, M.D., F.M.H. (Zagreb, Croatia: Agencija za komercijalnu djelatnost, March 2000), pp. 255-267.

[24] Simunovic, Z. "Pain and Practical Aspects of its Management." In *Lasers in Medicine and Dentistry: Basic Science and Up-to-Date Clinical Application of Low Energy-Level Laser Therapy—LLLT* edited by Zlatko Simunovic, M.D., F.M.H. (Zagreb, Croatia: Agencija za komercijalnu djelatnost, March 2000), pp. 269-301.

[25] Simunovic, Z. "Neurology." In *Lasers in Medicine and Dentistry: Basic Science and Up-to-Date Clinical Application of Low Energy-Level Laser Therapy—LLLT* edited by Zlatko

Simunovic, M.D., F.M.H. (Zagreb, Croatia: Agencija za komercijalnu djelatnost, March 2000), pp. 303-308.

[26] Rochkind, S. "Laser Therapy in the Treatment of Peripheral Nerve and Spinal Cord Injuries." In *Lasers in Medicine and Dentistry: Basic Science and Up-to-Date Clinical Application of Low Energy-Level Laser Therapy—LLLT* edited by Zlatko Simunovic, M.D., F.M.H. (Zagreb, Croatia: Agencija za komercijalnu djelatnost, March 2000), pp. 309-318.

[27] Simunovic, Z. "Sport Injuries." In *Lasers in Medicine and Dentistry: Basic Science and Up-to-Date Clinical Application of Low Energy-Level Laser Therapy—LLLT* edited by Zlatko Simunovic, M.D., F.M.H. (Zagreb, Croatia: Agencija za komercijalnu djelatnost, March 2000), pp. 319-328.

[28] Ortutay, J. and Barabas, K. "Laser Therapy in Rheumatology." In *Lasers in Medicine and Dentistry: Basic Science and Up-to-Date Clinical Application of Low Energy-Level Laser Therapy—LLLT* edited by Zlatko Simunovic, M.D., F.M.H. (Zagreb, Croatia: Agencija za komercijalnu djelatnost, March 2000), pp. 329-344.

[29] Simunovic, Z. "Laser Therapy in the Diseases of Ear, Nose and Throat." In *Lasers in Medicine and Dentistry: Basic Science and Up-to-Date Clinical Application of Low Energy-Level Laser Therapy—LLLT* edited by Zlatko Simunovic, M.D., F.M.H. (Zagreb, Croatia: Agencija za komercijalnu djelatnost, March 2000), pp. 381-383.

[30] Wilden, L. "The Effect of Low Level Laser Light on Inner Ear Diseases." In *Lasers in Medicine and Dentistry: Basic Science and Up-to-Date Clinical Application of Low Energy-Level Laser Therapy—LLLT* edited by Zlatko Simunovic, M.D., F.M.H. (Zagreb, Croatia: Agencija za komercijalnu djelatnost, March 2000), pp. 403-409.

[31] Strada, G.R.; Gadda, F.; Favini, P.; Baccalin, A.; Casu, M.; Dell'Orto, P. "Semi-conductor Laser Rays Therapy for the Treatment of Abacterial Chronic Prostatitis, Induratio Penis Plastica and Urethral Stenosis." In *Lasers in Medicine and*

Dentistry: Basic Science and Up-to-Date Clinical Application of Low Energy-Level Laser Therapy—LLLT edited by Zlatko Simunovic, M.D., F.M.H. (Zagreb, Croatia: Agencija za komercijalnu djelatnost, March 2000), pp. 413-418.

[32] Simunovic, Z. "Gynecology." In *Lasers in Medicine and Dentistry: Basic Science and Up-to-Date Clinical Application of Low Energy-Level Laser Therapy—LLLT* edited by Zlatko Simunovic, M.D., F.M.H. (Zagreb, Croatia: Agencija za komercijalnu djelatnost, March 2000), pp. 419-422.

[33] Simunovic, Z. and Trobonjaca,T. "Paediatarics." In *Lasers in Medicine and Dentistry: Basic Science and Up-to-Date Clinical Application of Low Energy-Level Laser Therapy—LLLT* edited by Zlatko Simunovic, M.D., F.M.H. (Zagreb, Croatia: Agencija za komercijalnu djelatnost, March 2000), pp. 423-426.

[34] Simunovic, Z. "Laser Therapy in the Gastrointestinal Diseases." In *Lasers in Medicine and Dentistry: Basic Science and Up-to-Date Clinical Application of Low Energy-Level Laser Therapy—LLLT* edited by Zlatko Simunovic, M.D., F.M.H. (Zagreb, Croatia: Agencija za komercijalnu djelatnost, March 2000), p. 435.

[35] Mikhailov, V.A. "Results of Clinical Study of Use of Low Level Laser Therapy (LLLT) for the Treatment of the Malignant Tumors of a Gastro-intestinal System." In *Lasers in Medicine and Dentistry: Basic Science and Up-to-Date Clinical Application of Low Energy-Level Laser Therapy—LLLT* edited by Zlatko Simunovic, M.D., F.M.H. (Zagreb, Croatia: Agencija za komercijalnu djelatnost, March 2000), pp. 437-454.

[36] Simunovic, Z. and Simunovic, K. "Lasers in Dentistry." In *Lasers in Medicine and Dentistry: Basic Science and Up-to-Date Clinical Application of Low Energy-Level Laser Therapy—LLLT* edited by Zlatko Simunovic, M.D., F.M.H. (Zagreb, Croatia: Agencija za komercijalnu djelatnost, March 2000), pp. 477-492.

[37] Pontinen, P.J. "Laseracupuncture." In *Lasers in Medicine and Dentistry: Basic Science and Up-to-Date Clinical Application of Low Energy-Level Laser Therapy—LLLT* edited by Zlatko

Simunovic, M.D., F.M.H. (Zagreb, Croatia: Agencija za komercijalnu djelatnost, March 2000), pp. 455-475.

[38] Wong, E et al. "Successful Management of Female Office Workers with 'Repetitive Stress Injury' or 'Carpal Tunnel Syndrome'by a New Treatment Modality—Application of Low Level Laser." *International Journal of Clinical Pharmacology and Therapy* 33(4): 208-211, 1995.

[39] Weintraub, M.D. "Noninvasive Laser Neurolysis in Carpal Tunnel Syndrome." *Muscle and Nerve* 20(8): 1029-1031,1997.

[40] Yu, Ishiban, M.D. "Low Power Laser Therapy for Carpal Tunnel Syndrome." *Proceedings of the International Congress of the International Society of Laser Surgery and Medicine,* Bangkok, 1993, p. 197.

[41] Naeser, M.A. "Treatment of Carpal Tunnel Syndrome: Research and Clinical Studies with Laser Acupuncture and Microamps TENS." *Proceedings of the 2nd Congress of the World Association for Laser Therapy,* Kansas City, Missouri, September 1998; pp. 145-146.

[42] Eduardo, C.C.P.; Marques, M.M.; Matson, E.; Pereira, A.N. "Effect of Low-power Laser Irradiation on Cell Growth and Procollagen Synthesis of Cultured Fibroblasts." *Lasers in Surgery and Medicine* 31(4):263-267, 2002.

[43] Nandakumar, K.; Obika, H.; Ooie, T.; Shinozaki, T.; Utsumi, A.; Yano, T. "Inhibition of bacterial attachment by pulsed Nd:YAG laser irradiations: An *in vitro* study using marine biofilm-forming bacterium *Pseudoalteromonas carrageenovora.*" *Biotechnology and Bioengineering* 80(5):552-558, December 5, 2002.

[44] Hosokawa, T.; Itoh, T.; Maegawa, Y.; Nishi, M.; Yaegashi, K. "Effects of Near-infrared Low-level Laser Irradiation on Microcirculation." *Lasers in Surgery and Medicine* 27(5): 427-437, 2000.

[45] Borisenko, G.G.; Kazarinov, K.D.; Osipov, A.N.; Vladimirov, Y.U.A. "Photochemical Reactions of Nitrosyl Hemoglobin During Exposure to Low-power Laser Irradiation." *Biochemistry (Moscow Journal)* 62(6):661-666, June 1997.

[46] University of Texas Houston Medical Center, news release, October 12, 1998, website http://www..uth.tmc.edu/uthorgs/ pubaffairs/news/releases/muradnobel.html

[47] Birngruber, R; Chaudhry, H.; Gregory, K.; Kochevar, I.; Lynch, M.; Schomacker, K. "Relaxation of Vascular Smooth Muscle Induced by Low-power Laser Radiation." *Photochemistry and Photobiologyk* 62(6):661-669, November1993.

[48] *The Journal of Clinical Laser Medicine & Surgery*, V-18, #6 2000

[49] Heinze, G.; Pernerstorfer-Schon, H.; Schindl, A.; Schindl, L.; Schindl, M. "Systemic Effects of Low-intensity Laser Irradiation on Skin Microcirculation in Patients with Diabetic Microangiopathy." *Microvascular Research* 64(2j):240, September 2002.

[50] Walsh, L. J., "The Current Status of Low Level Laser Therapy in Dentistry, Part 1.: Soft Tissue Applications." Aust. *Dent. J.,* 42(5): 247-254, 1997.

[51] Fini, M.; Giardino, R.; Giavaresi, G.; Guzzardella, G.A.; Torricelli, P. "Laser Stimulation on Bone Defect Healing: An *in vitro* Study." *Lasers in Medical Science* 17(3):216-220, 2002.

[52] Khadra, M.; Ronald, H.J.; Lyngstadaas, S.P.; Ellingsen, J.E., Haanaes, H. R.; "Low-level Laser Therapy Stimulates Bone-implant Interaction: an Experimental Study in Rabbits." *Clin. Oral. Implants Res.*, 15(3):325-232, June, 2004.

[53] Khadra, M.; Kasem, N.; Haanaes, H. R.; Ellingsen, J. E., Lyngstadaas, S. P., "Enhancement of Bone Formation in Rat Calvarial Bone Defects Using Low-level Laser Therapy." *Oral Surg. Oral Med. Pathol. Radiol. Endod.* 97(6):693-700, June, 2004.

[54] Pinheiro, A. L.; Limereira Junior, F. A.; Gerbi, M. E.; Ramalho, L. M.; Marzola, C.; Ponzi, E. A.; Soares, A. O.; De Carvalho, L.C.; Lima, H.C.; Concalves, T. O., "Effect of 830-nm Laser Light on the Repair of Bone Defects Grafted with Inorganic Bovine Bone and Decalcified Cortical Osseus Membrane." *J. Clin. Laser Med. Surg.* 21(5):301-306, October, 2003.

[55] Pinheiro, A. L.; Limereira, Jr., F. A.; Gerbi, M. E.; Ramalho, L. M.; Marzola, C.; Ponzi, E. A., "Effect of Low Level Laser Therapy on the Repair of Bone Defects Grafted with Inorganic Bovine Bone." Braz. *Dent. J.* 14(3):177-181, March, 2004.

[56] Pinheiro, A. L.; Oliveira, M. G., Martins, P. P., Ramalho, L. M., Matos de Oliveira, M. A., "Biostimulatory Effects of LLLT on Bone Regeneration." *J. Laser Therap.* 13 or http://garm.dyndns.org/whelan_lab/01/html/%20/pinheiro.html

[57] Barber, A.; Luger, J. E.; Karpf, A.; Salame, Kh.; Shlomi, B.; Kogan, G.; Nissan, M.; Alon, M.; Rochkind, S.; "Advances in Laser Therapy for Bone Repair." *J. Laser Therap.* 13

[58] Ozawa, Y.; Shimizu, N.; Kariya, G.; Abiko, Y., "Low-energy Laser Irradiation Stimulates Bone Nodule Formation at Early Stages of Cell Culture in Rat Calvarial Cells." *Bone* 22(4):347-354, 1998.

[59] Cruz-Hofling, A.; Garavello Freitas, Z.; Baranauskas, I. B., "SEM and AFM Studies of Rat Injured Tibia After HeNe radiation."

[60] Gable, P.; Tuner, J., "Bone Stimulation by Low Level Laser: A Theoretical Model for the Effects." *Laser Partner Clinixperience*, July, 2003, www.laserpartner.org/lasp/wbe/en/2003/0067.htm

[61] Blay, A. "Effects of Low Laser Irradiation on the Implants Bone Integration Mechanism: *in vivo* Study." Dissertation, School of Dentistry, University of Sao Paulo, Brazil, 2001.

[62] Sousa, G.R.; Ribeiro, M.S.; Groth, E.B. "Bone Repair of the Periapical Lesions Treated or Not with Low Intensity Laser

(wavelength=904nm)." *Laser Surgical Medicine, Abstract Issue 2002.* Abstract 303.

[63] Walsh, L. J., "The Current Status of Low Level Laser Therapy in Dentistry, Part 2: Hard Tissue Applications." Aust. *Dent. J.*, 42(5): 302-306, October, 1997.

[64] Stuebinger, S.; Deppe, H.; Donath, K.; Kolk, A. "Peri-implant Bone Regeneration After Co2-Laser Treatment." Abstract #0829, The IADR/AADR/CADR 82nd General Session, March, 2004.

[65] Walsh, L. J., "The Current Status of Laser Applications in Dentistry." *Aust. Dent. J.*, 48(3): 146-155, September, 2003.

[66] *Laser in Dentistry: Revolution of Dental treatment in the New Millenium,* Proceedings of the 8th Congress on Lasers in Dentistry

[67] William Ralph Bennett, *Health and Low-Frequency Electromagnetic Fields*

[68] Laser World, Swedish Laser Medical Society, *The Laser Therapy - LLLT Internet Guide*

[69] Tuner, J. and Hode, L., "The Laser Therapy Literature" in *Laser Therapy: Clinical Practice & Scientific Background.* (Grangesberg, Sweden: Prima Books, 2002) pp.391-393.

[70] Brosseau, L.; Welch, V.; Wells, G.; DeBie, R.; Gam, A.; Harman, K.; Morin, M.; Shea, B.; Tugwell, P.; "Low Level Laser Therapy (Classes I, II and III) for Treating Osteoarthritis." *Cochrane Database Syst. Rev.2004*; (3):CD002046.

[71] Brosseau, L.; Welch, V.; Wells, G.; DeBie, R.; Gam, A.; Harman, K.; Morin, M.; Shea, B.; Tugwell, P.; "Low Level Laser Therapy (Classes I, II and III) in the Treatment of Rheumatoid Arthritis." *Cochrane Database Syst. Rev.2000*; (2):CD002049.

[72] Lindstrom, L., "The Light Stuff: Cold Laser Therapy is Joining the Injury Treatment Team." *Washington Post*, page HE01, February 17, 2004.

[73] Kaptchuk, Ted J. O.M.D., *The Web That Has No Weaver.* (Chicago: Congdon & Weed, 1983), pp. 4 &. 7.

[74] Connelly, Dianne M., Ph.D. *Traditional Acupuncture The Law of Five Elements.* (Columbia, Maryland: Traditional Acupuncture Institute, 1994, 2nd. Edition)

[75] Firebrace, P. *Acupuncture: Restoring the Body's Natural Healing Energy.* (New York City: Harmony Books, 1988), p. 57.

[76] Adams, Marilyn, Lac.N.D., "Traditional Chinese Medicine." *Integrative Health and Self Healing,* (Issue #5, December, 2003) p. 12-13.

[77] Kaptchuk pp. 358-359.

[78] Eisenberg, David, M.D., *Encounters With Qi Exploring Chinese Medicine.* (New York, New York: Penguin Books, 1987, 2nd. Edition) p. 28.

[79] http://nccam.nih.gov/clinicaltrials/

[80] Lytle, Larry, D.D.S., Ph.D., *Low Level Laser User's Manual.*

[81] Selye, Hans, M.D., *The Stress of Life.* (New York: McGraw Hill, 1956).

[82] *Stress Without Distress.* (New York, Philadelphia: J.D. Lippincott, Co. 1974).

[83] www.icnr.com/DentalDistress Syndrome/ DentalDistressSyndrome.html

[84] Platzer, Werner, The Color Atlas/Text of Human Anatomy, Vol. 3, Nervous System and Sensory Organs.

[85] Dr. Koichi Miura of Japan—book not yet translated into English.

[86] Penfield & Rasmussen, *Cerebral Cortex of Man.* MacMillan, Co.

[87] Price, Weston A., D.D.S., *Nutrition & Physical Degeneration.* (New Canaan, Connecticut: Keats Publishing, Inc., 1989, first published in 1945).

[88] Pottenger, Francis, Ph.D., *Pottenger's Cats*. (Available from the Price-Pottenger Nutrition Foundation, phone-[619] 574-7763).

[89] Dr. Koichi Miura of Japan—book not yet translated into English.

[90] Fonder, A.C.: "The Dental Distress Syndrome (Quantified)." *Quantum Medicine*, Vol. 1, No. 1, 1988.

[91] Tuner, J.; Christensen, P. H., Low level laser in dentistry, posted on www.laser.nu/lllt/Laser_therapy_%20in_dentistry.htm.

[92] Tuner, J. and Hode, L., "Dental Laser Therapy" in *Laser Therapy: Clinical Practice & Scientific Background*. (Grangesberg, Sweden: Prima Book, 2002) pp.201-238.

[93] Tuner, J. and Hode, L., "Dental Research" in *Laser Therapy: Clinical Practice & Scientific Background*. (Grangesberg, Sweden: Prima Book, 2002) pp.239-276.

[94] Irons, Edward, Ph.D. *Death Begins in the Colon*.

[95] Neira, A.; Arroyave, J.; Gutierrez, M.I.; Neira, R.; Ortiz, C.L.;Ramirez, H.; Sequeda, F.; Solarte, E. "Fat Liquefaction: Effect of Low-level Laser Energy on Adipose Tissue." *Plastic and Reconstructive Surgery* 110(3):912-925, September 1, 2002.

[96] Ataoglu, S.; Cevik, R.; Gur, A.; Karakoc, M.; Nas, K.; Sarac, J. "Effects of Low Power Laser and Low Dose Amitriptyline Therapy on Clinical Symptoms and Quality of Life in Fibromyalgia: a Single-blind, Placebo-controlled Trial." *Rheumatology International* 22(5):188-193, September 2002.

[97] Segundo, A.S.G., Low-intensity laser coupled with photosensitizer to reduce bacteria in root canals compared to chemical control. Dissertation (Professional Master's Degree "Lasers in Dentistry"), Nuclear and Energy Research Institute, School of Dentistry, University of São Paulo, São Paulo, Brazil, 2002.

[98] Eells, J. T., Henry, M. M., Summerfelt, P., Wong-Riley, M. T. T., Buchmann, E. V., Kane, M., Whelan, N. T. and Whelan, H. T., Therapeutic photobiomodulation for methanol-induced retinal toxicity, Proceedings of the National Academy of Science, Vol. 100(6): 3439-3444, March 18.

[99] http://www.usuhs.mil/nes/Anders5.htm#links

[100] Seaton ED, Charakida A, Mouser PE, Grace I, Clement RM, Chu AC., Pulsed-dye laser treatment for inflammatory acne vulgaris: randomised controlled trial. Lancet 362(9393):1347-52, October, 2003.

[101] Carati, C. T.; Anderson, S.N.; Gannon, B. J.; Piller, N. B., "Treatment of Postmastectomy Lymphedema With Low Level laser Therapy: a Double Blind Placebo-controlled Trial." *Cancer* 98(6): 1114-1122, September, 2003.

Delivered in a South Dakota prairie sod house by his father during the great depression, Dr Larry Lytle learned a valuable work ethic. The youngest of seven children, he developed an early need to learn and reach beyond what seemed possible, as well as an openness for better solutions , always keeping his "antenna up." He is a student among students-- a teacher among teachers with several degrees--- a keen mind with a passion to share his knowledge and wisdom to help others achieve greater happiness and better health. His life's goal is to *"make a difference in humankind"* and his book *Healing Light* will make a difference---one reader at a time---starting with you!

Dr. Lytle graduated from Chadron State College in 1956 with a Bachelor of Science Degree. He taught high school biology and coached basketball, after which, he returned to college and received his DDS Degree from the University of Nebraska in 1964. He then practiced dentistry in Rapid City, SD from 1964 to 1998. During his dental career, he earned Category II accredited status in the Academy of Laser Dentistry and was accredited in the American Academy of Cosmetic Dentistry. He also earned a Ph.D. in Nutrition in 1979 and provided nutritional consulting in conjunction with his dental practice. His general dental practice included cosmetic dentistry, laser dentistry, TMJ/TMD and nutrition. Dr. Lytle lectured and taught in his many areas of expertise in the US and other countries. In the area of Proprioception, he was the developer of direct laser-bonded splints, Miracle Bite Tabs, and Easy Adjust Proprioceptive Guides. He also developed patents for low level lasers and has published the Low Level Lasers User's Manual. He currently is a consultant for several companies in areas concerning low level lasers and balancing the Sympathetic/Parasympathetic Nervous System via proprioceptive feedback to the brain and conducts Day of Learning Seminars in those areas around the world.

Dr. Lytle's main purpose in life is to make a difference in humankind and he passionately believes that his book *Healing Light* is an important tool towards carrying out that goal.

Dr. Larry Lytle
4020 Sunset Drive
Rapid City, SD 57702
Phone (605) 342-5669
E-mail – lytle@pobox.com
Web Site - www.drlytle.com

A FREE Offer for You
From Dr. Larry Lytle

Low level Laser therapy is truly *tomorrow's health care - today!* If you would like to receive a free information packet about low level laser therapy and the products I have created, including a fabulous 15-minute video (on CD) introduction highlighting the benefits you can expect to receive, please complete and return the information request form below.

Yours in Good Health,

Dr. Larry Lytle

To obtain your free information package, complete ALL the information below and either fax this form to (605) 342-5739 or mail to Wowapi Publishing, Inc., 520 Kansas City Street, Ste 201, Rapid City, SD 57701. Please allow 2-3 weeks for delivery of your information package.

Name _____

Business Name _____

Address _____

City/State/Zip _____

Phone _____ Fax _____

Email Address _____

FOUR Special Offers For You From Dr. Larry Lytle

As my thanks to you for purchasing this book, and because I truly believe that you and your family will benefit greatly from the information I have to offer, I want to provide you with the opportunity to learn more about low level laser therapy, proprioception and the field of energy medicine. This page provides you the opportunity to continue your education **at significantly reduced prices!**

Offer #1 - $40 off *Low Level Laser Users Manual*

This one-of-a-kind manual lists over 180 different health problems and provides protocols
…including photographs and laser acupuncture points…for using low level laser therapy effectively for each. Normal retail is $140 -- this "must have" manual is **yours for only 100** plus shipping with this coupon.

Offer #2 - $75 off *Healing Light Video Seminar*

Containing over 6 hours of education and training, my *Healing Light Video Seminar* covers all the information in the "live" version plus some! This video training session covers the A-Z's of low level laser therapy, energy medicine and proprioception. Normal retail is $500 -- **yours for only $425** with this coupon.

Offer #3 - $75 off my Live *Healing Light Seminar*

Offered monthly across the US, call (605) 342-5669 or visit www.laserinformation. com/event for details. Normal retail is $500 — **yours for only $425** with this coupon.

Offer #4 - All 3 products above for only $900!

Save an extra $50 - for **a total savings of $240** when you order all 3 products.

To order, circle your choice(s) above, then complete ALL the information below and either fax this form to (605) 342-5739 or mail to Wowapi Publishing, Inc., 520 Kansas City Street, Ste 201, Rapid City, SD 57701. Please allow 2-3 weeks for confirmation of your order and/or delivery.

Name _____

Business Name _____

Address _____

Note for **credit card orders**, please use credit card billing address. If shipping address is different please include separate sheet.

City/State/Zip _____

Phone _____ Fax _____

Email Address _____

Payment Method ☐ Visa ☐ Master Card ☐ Amer. Exp. ☐Check

CC Number _____ Exp. Date _____

Signature _____ Date _____

Providing this information constitutes your permission for Wowapi Publishing, Inc. and associated businesses to contact you via mail, fax, email and phone regarding related information.

Printed in the United States
72332LV00004BD/1-144